The Divine Comedy

of the

Tech Sisterhood

by

Anat Deracine

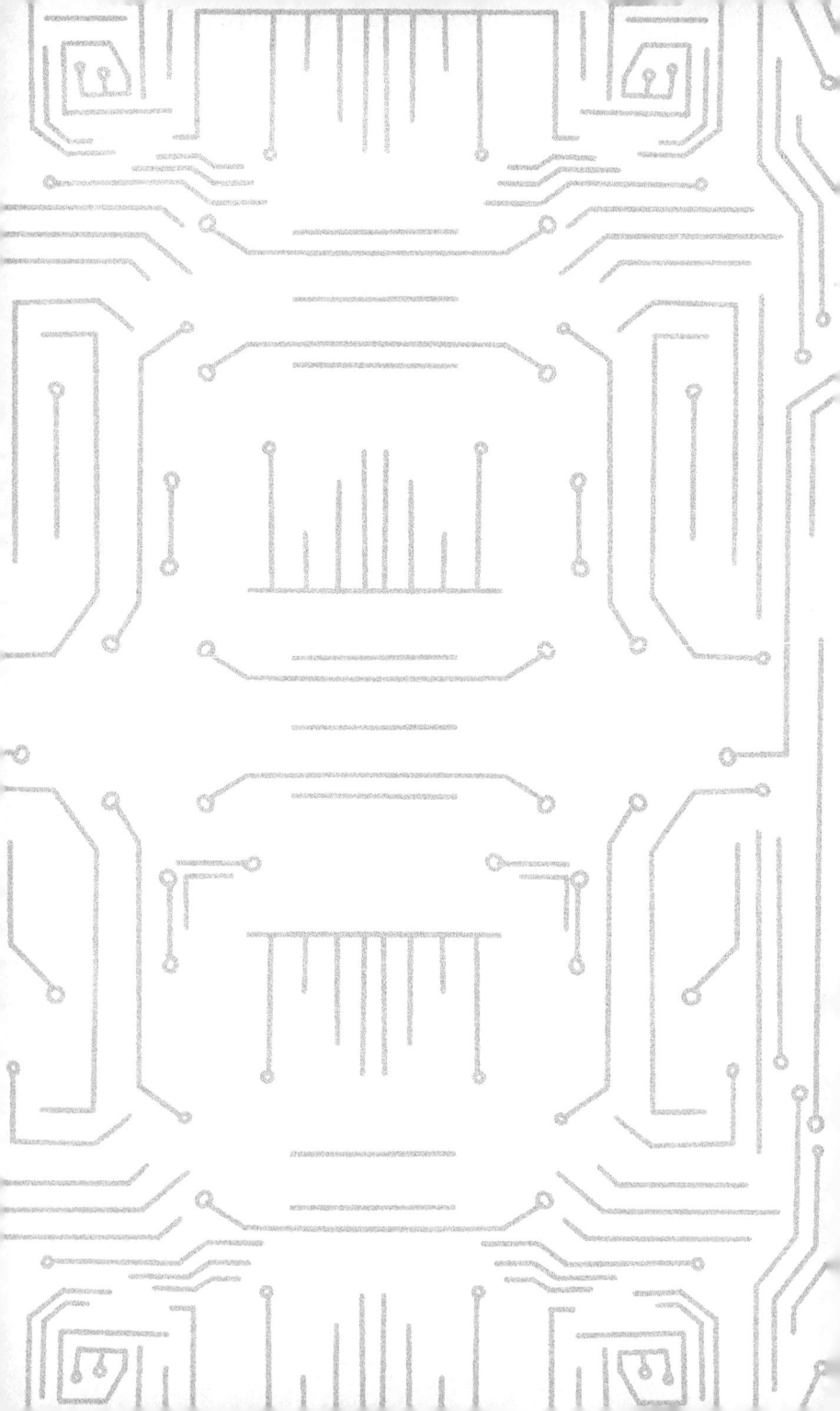

THE DIVINE COMEDY *of the* TECH SISTERHOOD

ANAT DERACINE

Copyright © 2025 by Anat Deracine
All rights reserved.
No part of this book may be reproduced in any form or by any electronic or mechanical means, including information storage and retrieval systems, without written permission from the author, except for the use of brief quotations in a book review.

Cover and interior art by Rena Violet

ISBN: 979-8-9908352-2-1

For every woman swimming against the current,
wondering why it's so hard…
breathe.
You're not alone.

1. The Divine Comedy of the Tech Sisterhood
(originally published in *Code Like A Girl*)

2. No Ocean for Young Women

3. Why do you believe you're a woman?
(originally published in *An Injustice!*)

4. Give a woman a mask, and she'll take you somewhere new
(originally published in *Mslexia*)

Afterword

The Divine Comedy of the Tech Sisterhood is a sobering yet empowering masterclass in the layers of challenges women face in even the most idealistic of male-dominated fields.

From the basics of imposter syndrome, unconscious bias, and gaslighting to far more complex themes of balancing identity and ambition, Deracine artfully guides by showing without telling. Watching the character vignettes play out is bittersweet and poignant, leaving readers who are still living it with a striking realization that this may be their own experience. The hope is that once the shock of recognition wears off, it can spur action.

This novella is therapeutic intervention in a book. It is important, a canary in the coal mine for change. Every woman in tech should read and re-read this.

— **Irene Chung,** Animator, Leadership Coach and former Google Manager

If I was in the business of adapting material for the screen, this would be my top choice. It's deeply cutting, raw in its honesty, nuanced in its storytelling, and above all, fearlessly bold. For anyone who's lived through tech's highs and lows, this novella is almost too real. It's nostalgic, hilariously honest, and a sharp reminder of why diversity truly matters.

— **Nancy Xu**, Product Manager / Design Manager / Film Producer

Right from the onset I was hooked in digesting the various characters involved in the story. As the scene was set the jigsaw of working in Tech emerged which I found fascinating. There is such an important message woven within the story about good leadership, not so good leadership and work expectations linked to specific characters. It is a must read for anyone working in tech.

— **Lorraine Warne**, Author of *The 7 Secrets of Great Leaders* and Founder & CEO of Cambridge Inner Game Leadership.

THE DIVINE COMEDY OF THE TECH SISTERHOOD

A letter from Virgil

IF YOU'RE LOOKING FOR inspirational wisdom about how leadership is more than a creative enterprise but a *calling* and there's nothing more rewarding than selflessly enabling other people's successes, this is not that.

This is a guide to some of the absurd dysfunctions and dynamics you will likely encounter if you embark on a career in tech, and end up in a leadership position with all the naive innocence and drive that *you* will be different, *you* will be a great leader, respected by all, and *you* will change things not just for yourself and your team but for the industry as a whole, moving the needle on behalf of all the other women out there who look up to you as a role model.

So, you believe in yourself and you've read *Lean In* and gone to Grace Hopper, and are armed with myriad certificates in leadership, influencing and negotiating skills. You're convinced you won't fail.

This is a story that you should come to when you've failed. When you're in pain, when all your creativity and emotional resilience has been depleted by the guy you fired who is now suing or stalking you, when you're considering quitting to have a baby, travel the world, write a book or maybe try your luck as a yoga instructor.

So is this a girl book? Self-help? Can a self-help book use the word fuck? Should I hide this book in a paper binding if I'm reading it on the subway? If you're asking yourself any of these questions, you're not a real engineer.

How'd that feel? Get used to it. Because any time you actually solve a problem, someone is going to say that what you do isn't really engineering. That you're not as sharp now that you're not writing code anymore.

This is a book about all the ways leading can crush your soul.

And about how to survive and thrive anyway.

PART ONE

INFERNO

A lot of the tech industry today is *fucked up*. Think of every horrible stereotype of software engineers you've ever seen on TV—hackers straight out of mom's basement who can't lift their eyes from your chest, socially-awkward twenty-something millionaires who get shit-faced in night clubs in San Francisco and convince you of the merits of polyamory, venture capitalists who refuse to shake hands with women for religious or cultural reasons—and somewhere in Silicon Valley, you will find it.

And that's not even scraping the surface.

The entrance to Hell is not a cave with the words *Abandon all hope*. It's the $200,000 signing bonus being offered to you by Meta, or the car from Microsoft, or the offer of free food from Google after you've scrimped and saved to get through college and are eating nothing but Ramen every night so you can chip away at your student debt.

Your journey begins with being rescued, and so for the first couple of years you struggle to be worthy of the gift you've been given, not recognizing the temperature increasing gradually around you to its boiling point.

Welcome to Hell.

ONE

ATHENA MEETS THE BROGRAMMERS

You've never been drunk in your life. You skipped a grade in high school, graduated summa cum laude from university, and so you're not yet twenty-one. Back home, in Kansas where your parents never went to college, or in India where your family lives in a tiny, cramped apartment, people are counting on you to make their sacrifices worthwhile.

You've never been an extrovert. You had very few friends in high school. People either bullied you for being a geek or stared at you when you demonstrated your intellect, as if you were Athena sprung directly from the head of Zeus and not a normal girl at all.

You've never learned how to dress yourself. You spent high-school and college studying, hiding in dark hoodies and unflattering jeans. You worked three jobs to put yourself through college, and one of them was washing dishes in the college cafeteria from where you came home with your shoes smelling like eggs.

You've never really dated, either because love was a distraction or sex was a sin, or the University of Waterloo where you studied Systems Engineering was a boring bomb shelter and your body went into hibernation for four Canadian winters. So you don't know when boys are flirting with you or just being friendly or treating you like a kid sister because all of it amounts to more attention being paid to you than you're really comfortable with anyway.

So when the boys invite you to their parties, you feel like you've arrived. You'll drink beer with them even though it tastes like stale yoghurt or pretend to like the really expensive whisky that brings tears to your eyes. You'll be their anointed Queen, singing and whooping as the world swims before your eyes, and taking the proffered Adderall to stay awake until 2 AM. They all claim that they see you as a friend and a sister, nothing more, it's all safe, but something tells you they want something from you even if it's only validation that they're cool enough to have a girl with them.

"Why do women not want to join tech?" one of them asks you. "Aren't you having fun? Isn't this fun?"

"You're not like other girls," the quiet one says when you're alone discussing the benefits of one API over another. "They're stupid or irrational. You're so smart."

"Let's go on a road-trip!" says the ringleader, your manager. "We'll be driving across the country as a team-bonding exercise. We'll go barhopping in the mountains, drink Red Bull and vodka by the campfire in the great plains, and Kerouac our way to Manhattan!"

Instinct tells you this is a *terrible* idea. You'll be the only girl among a group of drunk guys, stuck in a small, smelly vehicle. You'll have to either pay for your own room or share a room with them, when even the idea of a co-ed dorm had you shrinking in revulsion. You don't know if you'll be on your period.

But you can't say any of this. So you take your manager aside and tell him privately, "I feel weird doing this. I'm the only girl in the group."

He says, "Nobody's forcing you to do anything. It's just a team-bonding thing, and team culture is really important. You should have the opportunity."

What you hear is, *We need to talk about your flair.*

"If it's the sharing of a space that's a problem, you can drive alongside us separately and we'll catch up at the rest-stops. But you should know, the team really respects you as an engineer. They don't even really see that you're a girl."

You do what's necessary, because that's what you've always done. You toss away your lipsticks and your few skirts, you give up on makeup, you wear your hair in a messy, natural ponytail, and kill your inner anima not with anger but with rational detachment.

You join the brotherhood. You become an engineer.

TWO

IMPOSTER SYNDROME

"Sweet, everyone's here. Let's get started. Can you take notes?"

Since he doesn't know you or use your name, you look around to see if he really meant to ask you. Why you? He smiles encouragingly when you give the *Who, me?* face.

And then moves on to his content.

Why you? You look around. Oh, right. You're the woman.

Your ears are burning and you can barely pay attention to what anyone is saying. You're torn between taking notes because you're worried that if you don't you'll not be a team-player, and defiantly taking truly shitty notes so that he will never ask you again.

You spend most of the discussion calming yourself enough to ask him later, "Why'd you ask me to take notes? I couldn't take notes and participate at the same time."

"Oh, I'm so sorry," he says, giving a nervous laugh. "I thought you were the admin."

It happens so often that you're not even surprised. It doesn't matter that you're wearing the dingiest gray T-shirt you can stomach, or that you purposely failed to comb your hair this morning.

It's always an honest mistake. Interview candidates who refuse to answer your questions because they're interviewing for an engineering job and want to question your credentials first to know if you have the right to interview them. Sometimes you get mistaken for that other woman on your floor, or a shy new manager comes up to you and asks if you're the HR representative.

"I'm the other woman," says Sheila, laughing. "I never thought I'd find myself saying those words to introduce myself."

"Does it bother you?"

"It used to," Sheila says, her eyes glinting with mischief. "These days I make it work for me. Admins have all the power anyway."

You watch her at work and it makes you laugh. She listens to confidential HR issues never meant for her ears, schedules meetings with a dictatorial hand that always meets her needs before anyone else's, and her meeting notes highlight who is and isn't wasting their time in idle argument.

5 min: Jian presented the proposal with cost tradeoffs.

20 min: Ed said something about the performance characteristics of some new hardware.

5 min: Team agreed to Jian's proposal.

You start taking meeting notes too. You're a fast typist, so you collect a ruthless transcript.

Sheila: What we've noticed about iOS development is —

Dan: Don't even get me started on iOS. Do you remember the xkcd where... something something, who even cares.

Sheila: *innocent smile* Oh, I stopped taking notes because I was waiting to finish my sentence. Sorry!

Sheila's always smiling, but if you cross her she will get like a blade and cut you.

Sheila is who you're going to be when you grow up, if you don't get fired first.

THREE

ALPHA

"You've been doing tremendous work," Kris says. "*Tremendous.*"

You nod, because you've got to fight the urge to laugh at the way everything Kris says comes out sounding like dialogue in an action movie. Fake and bombastic, like his slicked-back hair and his slipping accent and his power-poses and the fact that his name isn't even Kris, it's Krishna.

"Since you've done all the work anyway, I was thinking, how would you like to be there when I present this to our VC?"

You wonder for a moment why you can't be the one to present, but there's probably a good reason, so you don't bother asking.

"Perfect. Make sure you suit up. I know it's not really *done* to dress up in the Valley, but the key to getting people to give you money is to look like you don't need it."

That. That right there is the kind of thing Kris says all the time, thinking he's being helpful, and it makes you think of all the horrible frat boys and douche-canoes in the Marina who add in clauses to their prenups against their wives gaining more than five pounds. Kris talks fast, dresses sharp, and annoys the *fuck* out of you just by existing.

You dress up.

Kris does his thing, says all the phrases you hoped to avoid hearing by never going to business school. You do a mental tally.

"Just to set expectations," five times.

"Strategic opportunity," sixteen times.

"Network effect," ten times.

"Cloud computing," twenty-five times.

Maybe it's a good thing you're not presenting. You wouldn't be able to keep a straight face. Kris talks about everything the product will do, and you wonder how the VCs don't notice that he only ever speaks in future tense.

"You look skeptical," says the VC, turning to you. "What are your concerns?"

Kris looks as if he's about to explode. You're not quite sure you'll survive this moment, never mind this day.

"The timeline is aggressive," you say, as reassuringly as you can. "We can't handle a million user signups without at least three months of work on production stability."

The VC nods and turns to Kris. "All right. Let's sync up again in three months so we can look at the reliability numbers."

For a long time after the meeting, Kris says nothing. You wonder if you should start looking for other jobs. Three months later, he takes you aside.

"You're still doing tremendous work," he says, giving you a smile that doesn't reach his eyes. "I just don't think it's a good idea for you to come to the follow-up meeting. There's a way these things are done, and it's not good for VCs or folks on the ground to see internal dissent. I need to know that even if you disagree with me, you'll do it in private. In public we've got to be a united front, and right now I can't trust that you'll fall in line. You understand, don't you?"

While you think of how to respond, he laughs and says, "Besides, I know how much you hate all that biz speak. I'm glad to provide you with air cover so you can be heads-down doing what you love."

You understand. You're a live wire, unpredictable and fiery, and your energy is best channeled into code. Looks like there's only room for one Alpha around here.

FOUR

KALI IS NOBODY'S VICTIM

"I need to talk to you," Ralph says.

"We're meeting tomorrow, aren't we?"

"I need to talk to you right now, about the performance feedback you wrote for me."

Alarm bells go off in your head. Ralph has the stormy look you last saw on the taxi driver who felt you didn't tip him well enough. It's the look of righteous indignation that, if you were not highly-educated professionals, could result in violence. "Let's talk about it tomorrow," you insist. "I've got a meeting now I need to go to."

"This will only take a minute. We could have finished talking about it already instead of arguing about scheduling."

You find an empty room so you're not having a messy confrontation in the hallway, even though your inner voice is laughing at you, saying, *Because the correct way to respond to an insistent and angry man is to be alone with him?*

But Ralph is all of 130 pounds, and you could snap his neck if you needed to. It's unfortunate that you have to consider everyone around you in pugnacious terms, but you learned very early not to walk into a room of men smelling like prey.

"I just don't understand why you would write about this stuff in my performance feedback," Ralph says. "I thought we were friends."

"What stuff?"

"You said I get easily frustrated and angry and that makes people not want to work with me. Why couldn't you just say that to my face instead of putting it on the record?"

You blink a few times because you can't believe that's not a rhetorical question.

"Look, we all have areas for development," you say in your most soothing voice. "My issue is having the confidence to set my boundaries—" *and this conversation is a clear indication of backsliding,* "—and yours is managing stress and frustration. It's not a big deal."

Ralph grabs you by the shoulders and shakes you.

"I DON'T HAVE ANGER MANAGEMENT ISSUES!"

When you were younger, you decided it was a good idea to fight the boys of your neighborhood for the right to have a picnic near their football field. The first time you felt a punch it caught you by surprise. The taste of blood, its copper-tang, was like a switch that flipped on a part of you that you didn't even know existed.

The blood is thrumming through your veins as you say, with deadly intent, "Your minute is up."

You leave Ralph sobbing in the meeting room. You know how the next few weeks will go, exaggerated kindness alternating with desperate demands for forgiveness that will border on ultimatums. *Do you still think I have anger-management issues? What the fuck? What do I have to do to prove myself to you?*

Ralph will quit when he can't bring himself to look at you anymore. You'll stay and spend one night a week at the gun range.

FIVE

GOSSIP GIRL

You don't like Julie. You really want to, because she's brilliant, and she's done absolutely nothing to warrant your dislike, and you're sick and tired of being the only girl on the team.

But does it have to be *Julie*?

She's gorgeous, in that careless Blake Lively way that means that she turns heads even when she's a complete mess. And she often is. A complete mess.

She's the kind of smart girl who never had to work for anything in her life, and she slouches back in her chair with her hands in her pockets and a bored expression on her face when you try to offer friendship or mentorship.

"Thanks," she drawls. "So far I think I'm okay. Everyone's so *nice*."

They really are nice to Julie in a way that they never were to you. Julie is invited to pub crawls and secret hangouts for anime aficionados. She can drink seven cocktails in a single night and still make everyone laugh. She speaks fluent Japanese, because her father hit it rich on some videogame and she spent her holidays surfing and getting cheap massages in hidden Pacific getaways.

Julie stops coming to your weekly mentorship coffee sessions. You don't even take it personally, because she doesn't need mentorship. People hang on her every word.

Three of the men on your team are in love with her. Each of them separately comes to you and asks for advice about "girl stuff."

"There's this girl I like, but I'm not sure if we're just friends or if there's something there."

"If you're hanging out with a girl every weekend, doing stuff together all the time, is she into you or using you?"

"There's this girl who really could have anything or anyone she wanted, and she's always talking about these loser guys she's going on dates with who can't even hold down a job."

This is how it ends: Julie chooses one of them, not to date but to fuck, when they're staying late in the office one night hacking on stuff, when Julie's bored or trying to prove a point to her latest ex—you never really hear the whole story.

What you do notice is the implosion of your team and the death rattle of your product. The gentle giant leaves first, his face white with shock and grief. The one who was spending his weekends with Julie, moving her stuff and driving her home after parties, he tailspins and you're left picking him up off the floor and holding his head over the toilet so he doesn't get fired. It's the last of the guys, Kyle, the quiet one, the chosen one, who really leaves you hating Julie with the fire of a supernova.

"Women are fucking manipulative," Kyle says, "Can't trust a word that comes out of their mouths."

"Generalizing from a single data point, are we?" you ask, not meeting his eyes. "Thought men had more analytical rigor than that."

Kyle freezes, and then laughs until tears flow down his cheeks.

"I'm sorry," he says. "I'm so sorry for everything."

He cries into his beer, and you realize you loved him all along but it's no use now. He still smells of Julie.

SIX

MANICURES AND MIMOSAS

T‍HEY SAY THAT MEN don't show stress the same way women do, that they bolt down their feelings and take their frustrations out on videogames. They get drunk. They get angry.

The people around you are stressed. You are too, but you do what most women do when they're stressed. You take care of everyone else.

You tell Aman to go home because he can't focus on anything but his sick kid, nor should he. You take on some of his work as well as Bob's, who is stuttering more than usual. It may have something to do with the way your manager Chris keeps asking for focus and results and clarity in daily status reports. Dave is somewhat impervious to stress because he got rich in an IPO, but even he seems to be unnaturally worried about bears entering his Tahoe cottage.

You notice that Emily is gaining weight around her middle, Freda has stopped going to the gym, Gary is subdued and isn't telling everyone stories about his tricks, probably because Harry complained that it wasn't work-appropriate conversation. Imran is being pressured into an arranged marriage, and Jane is pregnant and afraid to tell anyone, but her memory is failing and she's got deep bags under her eyes.

Really, you can go all the way down the alphabet on this one, to distract yourself from the fact that you're stressed out, and worse, you're feeling guilty about stressing out when everyone else has problems too.

Sarah is the first to snap, and you catch her crying in the ladies' room about Gary asking her not to waste his time asking him to explain things twice.

"Everyone's so loud. And *mean*," she says.

You're torn. A small, mean part of you delights in knowing that you're stronger than she is, and you want to tell her to get her shit together because this is a workplace, damn it, and we all have work to do.

"Do you want to go and get our nails done and talk about it?" you ask her instead.

She looks as if you just gave her the keys to heaven. It becomes something of a routine, where you both spend every other Sunday at the nail salon, bitching about everything that pisses you off, shedding frustration, cortisol and dead skin in favor of gel armor.

After the launch deadline passes, the two of you are commended for your work and given bonuses and bottles of wine.

Manicures and mimosas become a tradition. You wear your lacquers as a weapon, and you learn to put on your own oxygen mask before helping others.

SEVEN

ODE TO HILLARY

IN THE LAST FEW years, you've given everything you had to the work in front of you. You've dealt with your own fears that you were not equal to the task, you've given your best, you've handled a major launch and a production emergency, and you're clearly in line for a promotion or the leadership role on the team.

A-a-and it goes to someone else. Someone who isn't as technically competent as you, but is far more "likable" which, your manager Mike tells you, is an important characteristic in a leader to hold a team together in times of uncertainty. New Guy Nate is an expert at "influencing without authority" and uses "soft power." He's "very flexible in his leadership style, not directive and commanding."

If you don't dissolve into furious tears or head straight to the HR department to complain about sexism, you're a goddess among mortals.

You say, knowing that your voice is shaking, "Tell me more."

Tell me more about my likability.

Manager Mike sighs and says, "I'm sorry, I know you're disappointed. We were really hoping that you'd step up. We've been giving you feedback for a while about your leadership skills, but you haven't been very receptive."

"Tell me more."

"The other team members—obviously I can't tell you which ones without betraying their confidence—feel that you don't listen to them. You can be kind of aggressive, which silences and intimidates them. It's important that a business leader doesn't act like a military general, distant and commanding. It's important for a leader to be a good coach, to show vulnerability."

Suddenly it all makes sense. Mike is former military. He's six feet tall, has actual battle scars, and is a hundred times more intimidating and commanding than you will ever be. You spent the first year terrified of him and then learned that he valued you for speaking up so you adapted, learned from him, became his shadow self.

You are the mirror he can't stand to look at. He sees in you what you consciously added to your psyche by learning it from him, the steel required to hold your ground in the face of doubt, dissent, or condescension.

He continues, with a self-deprecating smile, "Now I know *I'm* not going to win any Miss Congeniality contests either, so it's a skill we both need to work on, and I'm happy to help you with it in any way I can. I really do believe in you."

Because you've perfected your mask, you smile and say, *Thank you*, and go for a walk to let the tears fall in private.

That evening you get a phone call from Rachel, who wants to talk. Rachel is trying to determine how to make her system integrate with one that was just acquired by the company. She met the lead of the acquired team to do the intelligent thing—ask naive questions. It's the Socratic approach, to question even the most fundamental assumptions of a software system, to begin from first principles instead of the party line.

"I asked him why they chose SQL and whether they had any concerns about scalability," Rachel says, laughing, "and he said that I should stop asking stupid questions and let real engineers do their job."

You close your eyes because the pain that underlies her laughter is just a little too much right now. You've got shit of your own to deal with.

"I told my manager what happened," Rachel said, "and of course he said it was unacceptable. But I was wondering, has something like this ever happened to you?"

"No," you say slowly, choking when you attempt to say anything more.

"That's what I thought," Rachel says, still laughing, her false cheer concealing her anger. "I realized that nobody ever pulls shit like that around you and I was wondering why that was. The dude and I are the same level, but he wouldn't treat me as an equal. But people even more senior to you are afraid of you sometimes."

Why do people not respect me? What do I have to do to prove I'm capable?

Why do people not like me? What do I have to do to prove I'm likable?

Insight strikes you like lightning from a goddess.

"We can't win," you tell Rachel. "We're either too emotional or too cold, too naive or too commanding, too junior or too set in our ways, too directive or too vulnerable… oh my God, we're Hillary Clinton."

At the end of the day, you'll both move on. Rachel will pick up a few things from you, like the body language and focused eye contact of a military general who might snap your neck or cut you down with a word, and you'll learn a few things from her, like perfecting the weepy puppy look when you just need something done by someone who wants to be a hero right now, and you don't have time to worry about looking weak.

You'll learn these rules because you're engineers who know that even society is just a system that's engineered according to certain principles that can be understood, questioned, and refactored.

You'll move on because you're engineers who, at the end of the day, really fucking love the work you do.

EIGHT

THE VIRGIN AND THE WHORE

EVER SINCE THE JULIE debacle, you've tried very hard to befriend more women, so you're not the turdy bitch who doesn't have any female friends. In fact, you're even friends with Julie now, who is older and wiser and regrets the way things turned out. (You'll wait a few years to get her to admit her agency in "the way things turned out" but this is progress).

You're friends with your exes, and your exes' exes, which turns out to be a survival skill in the Valley where everyone really does know everyone else.

So Dolly is a fucking curveball.

Seriously? *Dolly*?

Dolly is a doe-eyed princess from Cupertino. She was born in America but has never left the Valley except to go to India, where her family owns a palatial five-acre estate. She's married and has two adorable children who are as good at their Kumon classes as they are naive—neither of them have ever seen or heard of Harry Potter, because reading fiction or watching movies would distract them from their studies.

Dolly doesn't understand why her Masters from Stanford University isn't helping her navigate the organization.

"If you want me to lead the team, you should put me in charge of it," she says, pouting. "At my last company I had an idea and I went straight to the CEO. He thought it was a good idea and gave me six engineers who would implement it."

"They're looking for emergent leadership," you explain. "People here don't care about what you've done in the past. They don't care about your degrees or titles. They care about what you're doing right now, how you're able to help them, challenge them and lead them."

Dolly rolls her eyes.

"That's not *fair*," Dolly says. "So I have to win a popularity contest and keep pushing my way in or I don't get to be a lead?"

"Tell me what you mean when you say, *get to be a lead*."

"People don't respect you unless you have Team Lead or Project Lead in your title. At my last company I was a CTO. Here I'm nothing. If I bring my ideas to the table nobody listens."

Dolly has no idea how right she is. Tech is ruthlessly democratic and you're more likely to have listeners if you're doing stand-up comedy about your failed company than if you have the words Director or CTO on your resume.

"You can build yourself up," you offer. "If you're bringing a controversial idea to the table, don't spring it on a room all at once. Take it to a few people privately, ask for their feedback, reconcile their perspectives, and then when you know you have a lot of support and momentum, bring it to the meeting as a problem they can help you solve. Put them in a position to help you, and don't go to the senate unless you know how the vote's going to go."

Dolly's doe eyes get virgin wide.

"That's evil," she says.

"How's that?"

"If you want something, you should just ask for it. Not play games. That's what I tell my children too. I'm an engineer, not a politician."

"Think of it as engineering the outcome that you want. Think ahead about what you want to say, how people might react, and how you can mitigate any concerns in your approach."

The pout gets even bigger now.

"But taking people aside to talk to them separately, that's divide and conquer. It's evil."

"It's just a backchannel. People use it all the time. Here, look."

You show her a contentious email thread on the team mailing list that you brought to resolution, by taking some people aside to understand their positions, taking others aside to calm them down, pulling in a manager to make sure you were not overstepping by making a decision, stating a solution on the main thread and then having five senior engineers back you up, ending the argument.

Dolly's eyes fill with tears.

"So is everyone backbiting about everyone else all the time?" she asks. "Now I'm afraid to say anything in email because someone else might see."

"*This isn't backbiting*," you say through your teeth. "You just have to assume that people talk to each other, all the time, about everything. You will never be able to control who talks to whom about what."

"At my old company," Dolly says, "we didn't allow people to have a chat room unless the manager was also in it. Even if it wasn't the rule, my team would never *backchannel*." She spits the word out as if it were a depraved act. "They would always come to me with any problems. They loved me and I loved them."

You have to engineer the outcome you want, you remind yourself. It's true not just for mailing lists and meetings. It's true for every conversation. There have only ever been two outcomes possible from this conversation, and one of them isn't looking likely.

"It sounds as if you really miss your old team and company," you say, keeping your face completely neutral.

"It was very clear who was in charge of what," Dolly says agreeably.

"What would you like to be in charge of here?"

"I want a team of ten people who will just do what I need them to do," Dolly says, her tone indicating that she is asking for rather less than she thinks she deserves. She doesn't seem to understand that in software engineering, a team of ten people who will follow orders without question is kind of like asking for a unicorn and a dinosaur and your ex-husband to throw you a surprise birthday party.

"What do you propose to do with those ten people?"

Dolly shrugs. "I just want to be in charge of a project. I'm at the point in my career where I care more about leadership, not coding."

Partly out of schadenfreude, but mostly because you're curious about human behavior, you send Dolly to have lunch with Julie, suggesting that they might like each other.

Julie shoves at you playfully in the ladies' room. "Fuck you very much," she says, but she's laughing. "Four years ago you should've fired me, but you didn't have the power. If you don't fire her now, I'm going to have her sent to the colonies. She can lead up a team in India or China where people just want to know the rules and do what they're told."

"Racist."

Julie grins. There's a ring on her finger that's decked with diamonds. She's more beautiful than ever.

NINE

MANSPLAINING

JESUS CHRIST, YOU'RE JUST trying to get some work done. You would *really* like to spend your day reading design docs, deeply understanding the architecture of a ten-year-old system that your team is trying to fix so your company doesn't get sued for missing its SLA.

But you've got a problem. In what's supposed to be your ordinary weekly chat, Mark says he's surprised that his last performance rating wasn't as high as he'd expected.

He's even more surprised (and you're not an idiot, when he says "surprised" you know he means pissed off) that he only heard about it after it was already in the system (i.e. he didn't get a chance to negotiate).

Mark's a solid performer. But ratings are not actually negotiations. They're statements of fact about work that was done and expectations that were met, missed or exceeded. You went through an arduous process of fact-finding before giving him that rating with the full consensus of every other manager.

But, as I said, Mark's a solid performer, so you go the extra mile. You dig up your data and research, which, because you're a rockstar manager, is more detailed than any other manager has done for any of their reports. You think that maybe Mark will understand if he sees how you saw everything he had done when you made the rating call.

You email him the information, apologizing for the fact that he was surprised by the rating. Mark's a mature guy, he replies politely by saying that he'd like to read and digest before discussing in person.

It's Friday night, 10 PM, and you're enjoying the feeling of satisfaction that comes from having read those design docs so you can enjoy your weekend in peace.

Ping!

Email from Mark to you and your manager. It's long, and it pulls apart every statement you made and argues the minutiae. *It isn't fair to generalize*, he says. He insists that you give at least two examples for every single statement.

Remember the director gave me a bonus for my work, he says, even though the only reason he got that bonus was because you negotiated it for him with said director.

On and on the email goes.

I agree that I tend to try to solve problems by myself and don't escalate them in a timely manner.

This can be seen as normal/OK by experienced managers. For example, I prefer not to know every little detail, especially the ones that are likely to be resolved.

So I see this is as part of aligning ourselves to each other—and not something that impacts my ability to perform at a high level.

In other words, Mark is correct, you are incorrect. Moreover, you are a micromanager. He wants to tell his side of the story, not just to you, but to your manager who is more trustworthy as an adjudicator here because you don't know how to do your fucking job.

Because you're a demon slayer, you don't give in to your anger at the mansplaining. You don't try to defend your decision with facts, like Mark's project was delayed, and customers started complaining, and you ran interference for months to give him air cover because he was really doing a good job, if not a spectacular one.

Mark's not in the mindset to hear facts if he's sending emails to his boss and his boss' boss at 10 PM on a Friday night.

So you do the smart thing—you go dancing with the girls and shake it off. You know that Mark was triggered emotionally. But until you can detach from your own anger, you won't understand why or solve the problem.

Towards the end of the weekend the reason for his behavior becomes obvious. You remember the only other time Mark has *ever* pulled some stunt like this. It was when the coworker he was competing with reported to your director, while he was stuck reporting to you. He didn't understand that the director had years of experience and could manage a low performer, while you got Mark because he was supposed to be easy.

Mark pays attention to titles because it's the only articulation of power he really understands. He doesn't like his place in the food chain because it's below you. He's got years of experience on you, and he thinks you're keeping him down.

That's why he included my boss in the email, you realize suddenly. *He wants to move up the hierarchy.*

The simplicity of the solution makes you laugh. Two birds, one stone. You don't give a shit if he reports to you or to your boss or to the freaking CEO. Ultimately, the only thing that matters is who can figure out that ten-year-old system's architecture and fix it, because that's why you're all here, isn't it?

You don't reply to the email. You're going to chat with your manager on Monday. Your manager who really values your work and wants you to be happy. You're going to say, "It would make me and Mark happier if Mark reported directly to you. I don't want reports who don't believe I'm fighting for them with everything I have."

You'll talk to Mark in person. Butter won't melt in your mouth. You'll smile and say, with quiet dignity, "I didn't reply to your email because there's no point debating facts if you're emotionally upset. Are you ready to discuss just the facts now?"

Whether he is or isn't, you're done. You don't have to prove yourself to him. You don't have to explain why you're always given the most challenging technical problems while others struggle to have their efforts recognized and argue about how they might have screwed up four things but they did five things right and so it should all cancel out in their favor. You're not here to debate and prove that you can have a vagina and still be rational.

You're here to get work done.

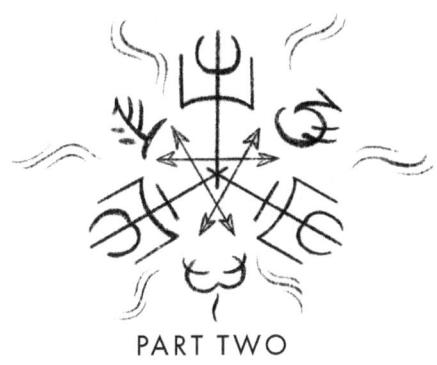

PART TWO

PURGATARIO

If you or a woman around you has never experienced any of the depressing crap from Part One, this section goes out to you.

You're a success story. You graduated top of your Ivy League class and had recruiters fawning over you and sending you gifts to get you to accept their offer. Founders of startups phoned you personally to talk about your career potential. You've been promoted a couple of times, you're sought out for mentorship by girls just out of undergrad who think you're a unicorn and headed for the C-Suite.

Life is great. You're a manager now. And your own boss isn't just openly gay and married, but has adopted daughters and lives in Noe Valley and understands work-life balance like nobody else in the world. Your team has more women than any other team in the company, and not one but *two* lesbians. You have no idea why everyone else complains about how hard it is to get diversity in the workplace.

Right now, you're probably glad none of the really awful stuff is happening to you. You're not aware of it yet, but things aren't as good as they could and should be, and when you do discover all the problems around you, you'll deny them, sweep them under a rug, bear up, and move on, or won't complain because you know life could be much, much worse.

You're part of the problem.

Congratulations, you're not in Hell. You're in Purgatory, which we all know is Hell-adjacent and filled with monsters.

It really, *really* sucks to be you right now, because you're going to feel so sideswiped when the shit hits the fan.

ONE

ARTEMIS TAKES NO PRISONERS

Ziv doesn't understand why his team of sixty engineers have been unable to launch anything in the last six months. What exactly are people doing all day if they aren't writing code? He's offered a cash bonus to anyone who takes on extra hours in production support and has promised to take all sixty engineers to Hawaii if they get a million daily users.

He lists off the management strategies he's tried while pacing up and down the long meeting room.

"I've created the right incentives, I've set an inspiring vision, I've motivated the problem by referring to the market need and what our competitors are doing, I've told the team I believe in them, I know every single one of them by name and I've made sure they're working on something aligned to their interests... I just don't understand, what more do they *want*?"

He's beautiful, you realize, in a way that speaks to your intellect as much as it does to your loins. A keen mind that chafes at being surrounded by mediocrity even as it tries genuinely to make room for others' limitations. You know exactly what that's like. You thought that you'd somehow be delivered directly out of university (which was clearly too easy for you both) into some sort of Avengers movie where everyone's superpower could be honed into solving an impossible problem. Instead, you're surrounded by regular people with banal problems. Housing costs, long commutes, getting kids into schools, and arguing about how impossible it is to go on vacation.

It helps that he's tall and lean, all sinew and restless limbs, and when he greets you it's with a boyish grin that makes you want to protect him from the world before his fire burns itself out.

"I want to make sure we're aligned," Ziv says. "I don't think there's a solution here. I've just been told I need to give you a chance."

"I like having a low bar to jump over," you say, sitting on a table and swinging your legs. "It's not condescending at all."

He's torn between finding your total lack of formality appealing and worrying about whether he's hired a child.

"Give me two weeks to find my feet," you say.

"Another thing," he says, with that adorably awkward hesitation. "We've never had a woman on the team, so people may not take kindly to it if they think I'm putting you in charge of them. You're an outsider, remember that, and some of these people have ten years of experience over you."

You bat your eyelashes at him, and he does a doubletake. It's clear nobody's ever tried flirting with him before. And why would they? His seemingly casual T-shirt says in stark black letters, *To what end*?

Everyone assumes that Artemis was a kind of Amazon, a woman whose shoulders were wider than an economy seat, who could chase lions out of the mountains and grapple with bulls.

Artemis was a huntress. *Stealthy*. Her step was light, her gaze was sharp, and her prey never saw her coming.

For the next two weeks you hang out in the common area and eavesdrop on water-cooler conversations. You have coffee with every single person on the team. You take the old guard out for drinks and loosen their tongues. You take the kids out dancing and hear about their love lives. You learn the important things.

Who cares about the project, who's only in it for the money, who's in it out of loyalty even though they don't believe, who's holding back progress by demanding all decisions go through them, who's hiding the facts from Ziv to stay in his good graces. You learn that they all worship him, but they think he's on crack if he believes this project is remotely feasible. They're all afraid to be honest with him, but they'll follow him into Hell.

Two weeks later you come back to Ziv with the decisions he needs to make. You ghost-write his talks for him, organize social events where you take potshots at each other in public so the kids can see that bantering and backtalk are perfectly all right. You quietly get fresh blood to infiltrate the old guard. Those who are made to leave thank you for the conversations that helped them realize that they'd be happier elsewhere. You work with the ones who remain to put an actual engineering plan together.

The project launches, and you get to go to Hawaii.

You're standing by Ziv's side at the beach in the moonlight, watching your team frolic in the waves with the indulgent and watchful eyes of parents protecting their brood of chicks.

"I'd heard you were good. I didn't know you were *that* good. It's like you have X-ray vision."

The buzz from the gin and tonic sends warm flushes through your skin. You've weathered a tough year together, and you know that there's something there between you that goes far beyond loyalty, but you can't call it love.

He shifts on his feet. Nausea roils in your gut.

"You know I'm bisexual, right?"

You don't even bother responding. The raised eyebrow says, *Obviously*.

"I used to be married. She was... a lot like you actually. Whip-smart, incisive and yet soft. A velvet knife." His voice falls to a whisper. "I can't do that again. I can't be with someone and wonder if they're controlling me. Your reins are gentle, but I know they're there. Guys are... well, they're just a lot more straightforward."

TWO

THE WOMAN BEHIND THE MAN

His name might once have been Samarth Padmanabhan, but he goes by Sam, and his keynote at the yearly leadership seminar he runs begins with his trademark line, "You have the *right* to be transformed out of obscurity."

Your heart skips a beat, because he's looking right at you.

It isn't something as prosaic as a crush. It's like having Steph Curry tell you that you've got great aim.

Sam continues his speech.

"The secret of my success is actually quite simple," he says. "Have only the best people on your team. I'd like to introduce——without whom I would never get anything done."

You've never blushed quite like this, not even at your senior prom. You're just a lowly sys admin, and not the kind that hacks into the CIA for fun, just the kind that fixes up parents' computers when they're fucked up with viruses.

You've heard rumors of other leaders who take the credit for their people's work, who provide zero air cover, who are always looking to find fault with their team so they can feel better about themselves. Horror stories.

Sam is unlike anyone else you've ever met. Even when he's telling you how you might have done something better (you've *never*, according to him, done anything *wrong*) his constructive feedback feels like a hug.

He's married, you've met his wife, and she's exactly what he deserves—gorgeous, smart, and sophisticated. You respect him more for his having chosen her over you, because your family Thanksgiving in one of the many Springfields, with two drunk parents and one farting grandfather, is not something you'd *ever* want him to see.

He's the rising tide that lifts all boats, and every year that he's received a raise you've got one too, and every time the most challenging opportunities come his way, he comes to you as helpless as a baby and says, "I can't do this without you."

You never stop to say, "You're right. You can't. So why aren't I in your seat?"

THREE

THE DISNEY PRINCESS

YOU COME BACK FROM a hard-earned vacation to find out that there's been an organizational shuffle. You have a new manager, which is disorienting enough, but one day your new manager shows up with a new team member and asks you two to be co-leads.

Let's call this one Belle. She has some sob-stories that she shares in perfect humble-brag humor. "At my old job they were *psychopaths* I tell you. I sometimes had panic attacks thinking about going in to work on Mondays."

It takes you two weeks to realize that she's really not doing anything *but* telling you sob stories. She's been through a lot—bad managers, misogynistic engineers, even the occasional harassment—and you're sympathetic, totes, but you're doing the lion's share of the work in this co-lead relationship, and she's not totally incompetent, but something isn't right here.

You bring it up as delicately as you can. "X has mentioned that their team needs Y. Can you handle it?"

"I don't want to step on your toes. You're so much better at it," says HumbleBrag Belle.

Suddenly you realize you're thirty years old but you're still that geek who just got asked to do the cheerleader's homework.

FOUR

FRIENDSHIP AMONG FAKERS

IT'S VERY IMPORTANT TO reach a data-driven decision.

But what does that mean? For which of the big decisions of your life have you ever looked at the facts?

You chide yourself for doubting. Data-driven decision-making would have helped you realize that Ziv was unavailable before you fell in love with him. What you really need is proper success criteria and monitoring for your blood pressure, your hormone levels, your attractions, your relationships, and your general mental and physical health. Come to think of it, you probably need to make a data-driven decision to assess whether the hundreds of dollars you're shelling on your therapist are even really worth it.

"Given this information, I could go either way," says Ziv. You struggle not to laugh at the double entendre, because he has his serious face on, the one that means that you are expected to put aside your complicated history and focus on the facts of this meeting. "What information is missing that would lead us to a resolution?"

What's missing isn't information. Whether or not this new product you're launching is going to be a game-changer or fizzle out like the fifty others you've tried before isn't something that can be predetermined by more accurate math.

Everyone has a theory, a way to squeeze the data to fit their beliefs. Harini in the corner knows that a new shade of blue isn't going to stop the product from being a solution to a zero'th world problem that will have no consumers in emerging markets. Harry the VC is listening intently because he believes that top-down solutions don't garner sufficient investment. Ziv, darling Ziv, who you know better than anyone, is rapping his fingers on the table. He's bored. He doesn't actually care anymore now that his product is launched and minting money, and the glaze in his eyes is because he's already packed for the White Party.

"Let's focus on the greatest impact."

"Let's look at the low-hanging fruit."

"Let's consider the alternatives—what's the cost of getting it wrong? Is any decision better than no decision at all?"

You have an idea: let's go party. *We're phoning it in, people,* you want to say. We're using the tired language of people who are too rich and comfortable to have any truly revolutionary ideas.

Ziv turns to you, now giving you his suspicious face. He knows you too well, knows your mask of disinterested objectivity for exactly what it is. "We haven't heard from you yet. What do you think?"

"I think that if we don't agree in the next ten minutes, we should cancel the project."

There is silence. Ziv gapes. It warms your heart to know that you can still surprise him, because his been-there-done-that attitude gets on your nerves.

The project is canceled in thirty seconds. The VC takes you aside afterwards. "It's a political game, you see.

I wanted to cancel the project, but I couldn't be the one to do it. Please do continue to be the voice of truth, the voice of dissent. Call out the Emperor when he has no clothes. You're headed for high places."

You stare at him. You want to say, "This company isn't going to last another year. Do you want to know that?"

You smile, say "Thank you," and send out your resume to competitors.

FIVE

CHARLIE'S ANGELS

HIS NAME WASN'T CHARLIE, it was Samarth, and he went by Sam, but there were more women in his team than in any other team in the company. People who joined came for Sam and didn't care what they worked on, didn't care if they ever got promoted. They got promoted anyway, as if Sam's ship was filled with pure helium and could lift a submarine out of the Mariana Trench without the slightest effort.

Sam was forty-five when he got bored. The telltale signs of executive boredom are even more obvious than the signs of grunt-stress. Sam started to dress better. He got divorced and remarried and did mushrooms at Burning Man. He got a nose job and took a few months off to go to Antarctica. He came back to talk about management philosophies, steer the ship, and confer his benedictions upon his legacy of leaders. He told self-deprecating jokes, insisted that passion was more important than loyalty, and boiled down his experience into pithy aphorisms that were quoted in the hallways.

It should have been a surprise to exactly no one when Sam left, and yet it caused more upheaval than you thought was possible. There were C-Suite tears and counter offers, interventions, and suggestions that Sam could continue to just dial in from Fiji every once in a while. They wouldn't take him off the payroll and so his name kept showing up in the system. A quit list was made, with the names of his most loyal followers, and the HR department made an outreach and retention program to target them.

You're on that list. So is Julie, whose red-rimmed eyes and frazzled hair are surer signs than any confession that she was in love with Sam. You wonder if they ever slept together.

Julie's full of smiles next week, and you know before she tells you that Sam's calling to her from beyond the bubble. He wants her to run his charitable foundation, and he's willing to pay her the salary she's making now. You try to tell Julie that running a charitable foundation is actually a step down for someone of her knowledge and abilities. It goes badly.

"You're still so naive, moshe," she says, using the nickname you hate. "In this industry you can't get anywhere without executive sponsorship. You don't know what it was like to be on this team before Sam was around. He just had to put a word in, and people got promoted. When he calls the CEO of another company, they make the time for him that day. It takes us three weeks to get an audience with their underlings. You don't even know the number of times Sam's gone to bat for us. He's called people out when they use terms like *bossy* to describe us, he meets with all the VPs to provide air cover so we're not subjected to bullshit, he takes the bullet for us when we fuck up. He's not just a great leader, he's an industry icon. This foundation of his isn't some stupid tax write-off. You should come with us. We're going to change the world."

Julie's gone. You're left with her people, some of Sam's people and yours, a large team of frightened college graduates who bring you chocolates and hand-written cards begging you not to leave them too.

You end up in a private meeting. There are two other people there: a frightened HR representative, and Kyle, who you haven't seen in a few years. You haven't really talked since the Julie debacle, and you assume it's because he doesn't want to be around the woman who saw him fall apart.

Kyle's a VP now. He looks different. Still the skinny, effeminate boy you remember, the one who blushed when he realized what really goes on in the San Francisco Armory. But there's a detachment to him now. His eyes aren't cold, but they aren't expressive either. His lips are still full and soft, but they're frozen in the Gioconda smile.

"You can leave us," Kyle says to the HR rep. She gazes at you in fear, because Kyle is a Big Shot now, and you're the little people she needs to protect.

You nod. There is no world where you could fear Kyle. He may have sealed off his heart, but there isn't even a shade of malice in him.

When the two of you are alone, Kyle says, "I'm supposed to sell you on the benefits of staying. You were on a quit ist when Sam left. I expected that. You ended up on another quit list when Julie left. I understand you were friends."

He says it without inflection or accusation, the journalist's tone. You can't resist digging at him.

"Julie and I have come a long way. Did you know we once dated the same guy?"

Blood rises to Kyle's ears.

"It's an interesting study of human behavior," you tell him. "Do those who are marginalized fight over scraps, or do they band together to amplify their voices?"

Kyle smiles. "There are other interesting studies. Like the perception of what people did versus what they actually did, and how that varies by gender. I see you're confused. Let me read you something. It's what Sam wrote last year, recommending that you get promoted. You're not supposed to hear this, which is why I sent the HR person away."

"I didn't get promoted."

"I know," Kyle says, and reads. He uses the "X" that your company's performance review process uses to mask people's names from the review board. It makes him sound even more dispassionate as he reads Sam's words. "One way that I know X is ready for a promotion is how she handled things while I was on leave. I was out for nearly three months, and when I came back, I found that lots of issues had come up while I was out, and everything had been handled thanks to X. Anything she wasn't able to handle was stuff only I could have done anyway. Feeling like my projects were in safe hands allowed me to step away and focus on other things."

Kyle looks at you. The mask is still on but there's sympathy in his gaze.

"It goes on," he says. "X is a leader because she listens intently and is very socially connected. She is able to gain the confidence of junior engineers and bring their thoughts to my attention. Ultimately I make better strategic decisions with her participation. X also helps with keeping the team cohesive and handling people's emotional issues, so I owe a part of my management award to her."

"*Stop.*"

Kyle does. For a long time, you sit there in silence. Kyle knows what this is, the time you need not to cry. Fuck Sam, for making everything you've ever done about him. Fuck Julie, for believing in Sam and making you believe in him. Fuck Kyle too for good measure.

"There are people in HR trying to decide your fate," Kyle says finally. "They know you've got more potential for leadership in your pinky finger than ninety percent of this company. But they think I'm the next Sam and you're the next Julie, because Sam was a cult of personality and Julie is the only woman they've ever known. But I'm not Sam, and you're not Julie."

Still, you say nothing.

"I don't want to be your Sam," Kyle says. "I know it would make things simpler for everyone else if we fit their mental model for the charismatic male lead in the movie and the woman who stands behind him, invisibly making him successful. But that's not what I want, because it would hold you back. You're my equal and my friend, and the only person I've ever known who's called me out when I was being an idiot."

It's this that finally makes the tears fall.

"I *could* protect you from the bullshit that goes on in exec meetings the way Sam protected me. The shit's going to hit the fan soon. There's a bust on the horizon. Do you want to be protected from it or do you want to grow up?"

SIX

THE MENTALIST

When Patrick Jane does it, it's entertaining, the colder aspects of his keen insight forgiven because he's grieving for his dead wife.

When Sherlock Holmes does it, it's scintillating, the knife-edge of his psychopathic curiosity forgiven because it's Benedict Cumberbatch, and God, those cheekbones.

When you do it, it's witchcraft.

You have a memory for faces so strong that you can pinpoint guest actors on *Castle* that you've seen once in *Supernatural* or *The L Word*, and it means you know *everyone* around you.

You recall not just when you saw a person last, but what context it was in, where they were sitting, in what room, and if pressed you can recall the entire conversation.

Speaking of conversations, you know how to steer them like a detective looking for clues, so you're always getting exactly the information you need out of people so quickly they feel glad to be useful in your investigation.

You read everything—rooms, faces, emotions, tensions—not to mention a vast and varied literature, so can draw even the shyest person out into a discussion of the latest sci-fi novel or hold your own in a discussion about database schemas or Greek philosophy.

You can recite the checklists for a software release process or an airplane takeoff.

You're probably a novelist, an Olympic weightlifter, a ballet dancer, or one of those people who posts pictures from your Parkour exercises at Athletic Playground.

You're a problem.

"How can you tell who is in a room just by *smell*?" Dave asks you, wide-eyed and squirming.

If you're a man and you reply with, "The scent of white musk is distinctive, but what's really telling is why you would choose to wear a fragrance selected to emulate the natural odor of a male deer attempting to mark his territory and find a mate," you'd get punched in the face.

When you reply with, "Our sense of smell is not just the most closely connected to memory, it's how we connect with our emotions and with other people," you get the response you had to know was coming, because you're just that kind of genius.

"That's fluffy psychological crap. I wish people would just engage with each other logically, you know, like *engineers*."

You want to tell him that Descartes was wrong, that identity is not a brain in a vat and it's not a box that might or might not contain Schrödinger's cat.

You want to tell Dave that what makes you good at your job is that you're firing on all fucking cylinders, mental, emotional and physical, and sometimes you're so alive it hurts.

It hurts to be alone in a room full of people who aren't even really *there*.

SEVEN

THE WOMAN WHISPERER

You hear so many voices you start to wonder if you're schizophrenic.

"My manager says I'm too aggressive. But I was right, and she wouldn't listen because I'm new."

"I don't want to be a lead. Have you seen the guys on the team? They're six feet tall and they stand together like a wall of thugs. I'm not going to get into a shouting match."

"I'm getting really sick of Hari. He talks all the time, never listens, and assumes that I agree with him because I didn't get a word in edgewise."

"I'm the only woman on my team. Where's everyone else?"

"I don't like my manager. He's pushy and moody and keeps telling me why I'm wrong and that I don't know what I want because I'm new. I spoke to HR and she said I should learn to work with him. Two other people have left my team. I want to leave too."

"I keep getting told to challenge myself and expand out of my comfort zone. But I don't want to take on the stress of being in the spotlight with everyone picking apart everything I say and do."

You've got an anonymous forum to support people with their struggles. Mentorship, recruiting, coaching, and efforts to improve life in the workplace make up nearly twenty percent of the week and fifty percent of your brain space.

Kyle sends you to Tel Aviv to speak with the company he's just acquired, and after your talk wide-eyed new grads from Eastern Europe ask you, "Is it true that in America women can get paid the same as men? I asked my professor and he said it was only in America that men and women perform the same. In this part of the world, men are better than women because they have better grades."

Your plan is simple. You have faith in the system, and if the problems people are facing persist, it's because they're not well-understood. The stories you're hearing will help you understand what needs to be done. You have no doubt that you can fix the problem once you know how. There's no system you can't deconstruct and re-engineer.

You and Kyle compare the salaries of your people. There's a gender gap, of course there is, because every company that claims to have fixed this adds a clause that says, "when adjusted for performance, men and women make equal pay."

You and Kyle know better. You know that performance is never evaluated fairly and so it's easy for people to justify saying that anyone can be a leader as long as they're assertive, confident risk-takers who lean into conflict and yet build consensus through soft power. That is, men only make more than women because they're performing better.

The social experiment goes like this. You and Kyle write evaluations for your people. When a man and a woman seem to you two to be performing equally, you use the same text for both. Kyle submits evaluations you wrote for his people in Silicon Valley. You submit evaluations Kyle wrote for the people in Israel.

The HR ratifications committee reads your evaluations. The chair sends you the following statement:

In a leadership position, there is an expectation of calm under fire, of hiding frustrations. Sierra seems to not be demonstrating sufficient diplomacy in her leadership style to warrant such a high rating. Rat.committee recommends manager consult with Kyle V—on writing more objective evaluations.

You watch the cloud pass over Kyle's features when you forward the report to him. His lips tremble, and your heart races to watch his fury. Your eyes meet.

Kyle walks away without a word. You have a new HR chair the next day.

EIGHT

THE BUTTERFLY EFFECT

MOHINI RESENTS THE RUMOR that she's the New Julie. She never even really knew Julie, but Julie's name echoes through the Valley as if it were her own Manderley.

"Julie was nothing without her man of the moment," Mohini declares, so much authority in her voice that the women's group around her just nods. "Julie didn't bother making friends with other women. Julie thought diversity efforts would stain her resume, as if she'd be lowering the bar, not raising others up."

More nods. All of this is probably true, but you don't like that it has to be this way. Julie was more oblivious than malicious, and if she never helped out the sisterhood it was because she had never felt the need of it herself.

"It's got to be some weird edge case failure of the Bechdel test," you mutter, "that there are fifteen women from every major tech company in the Bay, all gathered to talk about technology and we're gossiping about a woman who isn't even here."

Mohini's face darkens for a second, because she's been called out. Then she changes the subject gracefully. Later, she takes you aside and says, "I get that I crossed a line, but if you want to be part of the group in the future, it's important that you don't resort to public shaming, especially in front of my friends. You understand, don't you?"

You do.

"I just get triggered every time someone compares me to Julie. I mean, really, what do we even have in common? All women aren't the same, am I right? Hugs? There we go. We gotta stick together."

Mohini is not Julie. Mohini is a Hinjew from Edison, New Jersey. Mohini's parents run a restaurant. She grew up in poverty with four brothers, one of whom is serving time for dealing drugs. She's scratched and clawed her way into tech and has screaming matches with her mother in the office parking lot over her inability to find a husband or submit to an arranged marriage.

"Mohini reminds me of Julie," Kyle says at lunch, completely oblivious.

"Don't go there."

"They're different in one important way," Kyle continues, and only the slight deepening of his tone is any indication that this is critical information he's about to share. "The past didn't even really exist to Julie. For a while after… I didn't even try to talk to her. I was afraid she'd humiliate me publicly, make it so I'd never get a job anywhere. Careers can be destroyed so easily. All it takes is a photo. A forwarded email. A tweet."

His words are a warning, as if you needed one. Through the glass you watch as Mohini makes her daily rounds, asking people if they want to join her for lunch or a board game as a break. Her eyes fall on the two of you on the balcony, and her lips purse in disapproval.

Why? What does she think this is?

"But Julie could never be bothered with anything as petty as jealousy or vengeance," Kyle says, his emphasis on the name of the ghost the clearest sign of impending doom.

"That's not the kind of thing you can know about a person, or even about yourself, in advance," you say. "You never really know the kind of person you are until you're truly tested. I don't know, for instance, if I had a chance to confront someone who had really hurt me, whether I would take it, whether I'd hurt him back if it wasn't a crime."

Kyle's eyes grow distant. "But you know, don't you, what holds your power in check? What keeps you from abusing the position you already have in search of a higher one?"

His question leaves you unsettled. It's only three months later that you understand. Mohini has been amassing power. In the world of tech, where everything is discrete and measurable, the idea of amassing pure *power* is as alien as trying to chant spells to get code to compile. But Mohini has run a campaign, quietly and inexorably, created a platform for herself. It doesn't include you.

There are parties with VCs in SOMA lofts, fundraisers in the pop-up art galleries in Bernal, and back-rooms of the bars begun by hit-it-rich party boys are booked out weeks in advance. Mohini has become a sort of patroness of the geeks, able to discuss anything from the merits of Java over C++ to which indie band must be followed *now* before they get too bougie to bear. But you're off the guest list.

There's another marked difference between Mohini and Julie. Mohini never sleeps with the men in her set.

It's a gaggle of women, gay men, and awkward engineers who are so far on the autism spectrum that they attend her court and never say a single word, as if there were marks for participation.

None of this really matters to you. You do learn something from her though. There are ways to be a social nexus that have some unprecedented effects. Mohini is the People's Queen, the voice of a proletariat that didn't know it was disenfranchised and oppressed until she told them so. When she's feeling generous Mohini acts as the Intermediary, bridging the concerns between divided groups but still perpetuating their division.

"If you want a contact at X, a real human contact who'll actually get shit done for you and not send you into their process twilight zone, you gotta go through me," she announces irritably at a meeting, and the engineers gaze up at her in awe. She can procure anything. Cheap opera tickets, a branding agent for a team logo, front row seats to the baseball game, a way to bypass the line at the best restaurants.

"Don't be talking to my peeps without telling me!" she says with a friendly smile when she catches you at *her* restaurant. "Yo, Marco, you make this chica happy, whatever she wants. She's one of mine, you hear?"

You're so used to thinking of mobs and the mafia as a man's game that you never saw the Valley turn into a turf war beneath your feet.

Kyle needs to make his six-monthly global tour, something every VP in the Valley is encouraged to do in order to get a better perspective on tech developments in other countries and the needs of emerging markets.

He sends you an IM. *Who do I need to talk to while I'm out? What should I make sure to try? Where do you think I should go?*

I'm not some fucking yenta, you write back. Mostly you're annoyed because you're planning a trip to Peru and everyone you ask for advice tells you to talk to Mohini, who *really* knows how to travel.

At the team-leads meeting that day, Kyle grills you ruthlessly. It's not done with anger, and his tone is perfectly even, but he's asking more questions about more details than he should really care about. Something is very definitely up, but his face is a mask. At the end of the meeting, he says to Ryan in a way that tells you it's actually meant for your ears, "I hear your concern, but she can take it. And at the end of the day, I need everyone to have depth and substance. I don't want my leads to be air traffic control, simply forwarding questions to other people."

News of your excoriation ripples its way down the ranks, and Mohini shows up at your desk with your favorite tea. "I heard what went down," she says. "We haven't had a chance to catch up lately. I know you're probably feeling betrayed right now, but these guys, they can just turn on a dime. One minute you're their pet, the next minute you're out of favor. That's why we've gotta stick together, sisters before misters, am I right?"

You go home and spend three hours putting an itinerary together for Kyle. Because you know him better than Mohini does, and what happened today was the best gift anyone's ever given you.

NINE

THE DEADLIEST SIN

YOU'RE THE RISING STAR. You've got the most challenging projects. The best team members want to work with you, and in fact threaten to quit if you're not made happy. Stock refreshers and bonuses are rolling in until you're paying more in taxes than the 99% make as their yearly salary.

And then you don't get promoted, even though every single person around you believes you're more than competent. You know from the way people don't meet your eyes that there's a reason for this that they all know but aren't telling you. There's a widespread secret that you're not privy to. You're not invited to the leadership circle, even though you know more about the systems and the people than all of them put together.

What the fuck.

Kyle, bless his heart, tries to explain. He says, "I know you can handle direct feedback. You're the kind of person who can take anything without flinching. Every single lead loves working with you, but they've all said the same thing: you're exceptional, but you come across as junior in your approach. I'd like to work with you to find out why that is and fix it."

What the hell does that mean, *junior*? You've proven yourself over and over and yet again. In every design review, you're the one who asks the crucial questions. You've got more technical breadth and depth than the vast majority of engineers you work with, and you're not afraid of any challenge.

"Tell me more."

"I think it might help if you showed more vulnerability."

You spend a weekend thinking about this, fighting the waves of betrayal over the idea that Kyle wants you to put yourself in damsel-mode. *Et tu, Kyle?* And what does it even mean to show vulnerability? And can you really just cut yourself and bleed into the water and *not* get eaten by sharks? And is it even in you to be vulnerable, to be the weepy puppy asking for scraps at the leadership table instead of demanding your place at their side having earned it by merit?

Fuck, fuck, fuck.

You don't sleep for three days. You might be able to take anything without flinching, but you're not made of stone; you just have a cement mask, and the walls around your heart are thicker than Troy.

Maybe that's the problem. Maybe you're blind and unhearing of subtle messages because your heart's so closed off. *Show more vulnerability.* Fine. Suddenly your next step is clear.

You go to Leo. He's the boy genius who's likely going to be CEO of something and change the world before he manages to grow hair on his chest. His team doesn't just respect him, they love him with the fanaticism formerly reserved for Joseph Gordon-Levitt.

Oh, shut up with the explanations. You've got a crush on him. That's why you go to him and not to some leadership coach. You want to know if he's paid any attention to you, if he'll tell you differently, if he'll say, "Fuck this vulnerability shit. What I love about you is how assertive you are, how you've paved the way for women around you and changed everyone's perceptions with your exceptional competence."

"I'm trying to understand why people think I should show vulnerability," you tell Leo. "Do you have any guesses as to what might be going on?"

He asks for more information—how does your team function and relate to one another? What are the mentorship structures? What are common complaints? How do meetings run? How does performance management usually pan out?—And all the while you're thinking, *Yes, this is what I need, someone who really sees into the heart of things, who sees into my heart and can hold up a mirror.*

He gazes off to the side in contemplation and then says, "It sounds like the crux of the problem is that it's all about you. I mean, your team clearly respects you, but they're not stepping up. You've got a strong personality, so they get relegated to the shadows. Normally, I'd say you should ask more questions, but that's actually something you do really well. You ask a lot of questions, but maybe not the right ones. When you share things with others, it's always about your experiences and what you learned from them. When I was doing X, I learned Y. I know you don't mean to monopolize conversations, but you do."

You. Will. NOT. Flinch.

"Tell me more."

Leo sighs. "I'm trying to remember the conversation, but I don't have your memory. We may have been having a discussion about how to manage low performers, and you offered your opinion based on the one you'd handled. I mean, you were correct, so the information you provided was constructive, but I remember thinking, *everyone here isn't just more experienced than you, they*

TEACH classes on management. So you didn't really see your audience. I mean, dear God, woman," he laughs, "you're competent, but you don't have a monopoly on competence. It wouldn't kill you to show some humility."

There's really no charge to his words. He's bantering with the directness you love, treating you with the respect and frankness you yourself have demanded. If he were treating you as *junior* (thank you, Kyle) he would have coached you rather than told you. He cares enough about you to make your time together worthwhile, to give you exactly the information you need as soon as you need it.

You thank him for his time and go for a walk. Your hands are shaking and your knees are weak. You don't know if you'll fall apart. You don't have tears because you haven't cried since you decided that you wouldn't show weakness, but damn, it would feel amazing to fall apart right now, to cry and be hugged and told that you're not just respected, but loved.

It takes a full thirty minutes for even a single tear to fall. Why is it that you've handled being told you're incompetent, being passed over unfairly, being shoved into a wall for fuck's sake, but this hurts so badly you want to quit? You think of that leadership circle meeting in the afternoons, the one with no women in it, and you think about Leo sitting in the room telling Kyle over beer, "Yeah, I think she heard the feedback. She came to me and I told her straight out, she needs to come down off her high horse before anyone will take her seriously."

He wouldn't do that. But you hate him, and you hate everyone and everything right now.

Many people have tried to cut you down, but this is the first time someone has broken your heart.

PART THREE

PARADISIO

IT GETS WORSE BEFORE it gets better. They may call you Moshe, the one who rescues and is rescued out of the water, Moshe, the prophet who was raised as an Egyptian prince but never forgot that he was one of the enslaved Hebrews. Seer and savior, that is what you are called, for your unerring instinct and your ability to see deeply into everything and everyone, but you never see the wave coming and you can't save everyone when it comes.

Ten years in, they're all gone. It happened slowly, so slowly you didn't see. It didn't happen near you, it happened in other companies, other teams, other cities and countries, and so you didn't see.

Harassment. Microaggressions. Unequal pay. Slow promotions. Misogynistic language. Drinking culture. Useless HR. No HR. Burnout. Anxiety and depression. Pregnancy. No maternity leave. No privacy in the "collaborative" open offices. No financial support or recognition for outreach and diversity initiatives. Tone-policing from "allies" who insist that more people would get behind the cause if they weren't so emotional about it.

And these are just the things that a few people have been willing to talk about, and only to you, because they're worried about being doxxed.

Doxxing. That's a thing now, and so it's not bad enough that people leave their jobs, leave the *entire industry*, these women have to pick up and leave their homes because their personal addresses have been posted online and they're receiving death threats.

Companies throw money at the problem frantically, blaming the pipeline and questioning the quality of HBCUs, but as Kronda puts it, when your bucket is leaking half the water in it, do you (a) get a bigger hose and put more water in the bucket or (b) FIX THE LEAK?

The darkest hour is when Kyle gets up in front of all 400 of his people and says in his neutral, dispassionate way that we need to make the workforce more inclusive of diverse backgrounds, and there is a rumble of displeasure in the crowd.

One man stands up, adjusts his trousers with trembling fingers. Kyle doesn't see it, but you do: the banked rage.

"I don't understand," the guy says. "You say you want to bring more Blacks and Latinos and women. You want to bring affirmative action into tech, but we're engineers. We only make decisions based on the merits, and we only hire based on the merits. In Harvard admissions, Asians worked harder for their entire lives and scored hundreds of points higher on SATs, but we still could not be admitted. Now you want to bring that discrimination to tech and lower the bar for these people. When you talk about diversity, why does no one care about Asians?"

You know that this is going to turn into a shit storm. You don't know how yet, so you start messaging people in the room frantically. Who's this guy? What's his story? The answer reaches you through the grapevine, because people know now to give you the information you want without question. It's Hector, who could not get into Harvard or MIT and had to "settle" for Urbana-Champaign, and still bristles over the fact that everyone else on his team went to an Ivy-league school.

Kyle says with his patient tone, as if he were talking about a site outage rather than a triggering racial issue, "Rather than think about this in terms of one group versus others, I would encourage us to consider that not everyone has had the same opportunities, and that by applying our unconscious biases that favor confident, extroverted, well-dressed white men from big-name universities, we may be discriminating against people who—"

"THIS IS BULLSHIT!" cries Hector, and the room falls silent. "I worked for my entire life to get this job. You want to hire someone who doesn't even have a Computer Science degree to take my job?"

It's the new HR chair who steps in, with a bland recital about respect. She goes into some hiring profiles for the year, listing the number of Asians that *have* been hired, and the muttering gets louder. You know that this moment is why you're here, and so you walk up to the front of the room. Kyle steps back without a word, even though neither of you knows what you're going to say.

"I want to thank you, Hector, for raising an important issue that is clearly a concern for many people. I want to thank Kyle, as well, for creating a culture where we can raise tough issues about race, gender, and diversity and discuss them together."

The room is still rumbling, but the volume is lowering. Hector's fists are still clenched at his sides. Kyle is stiff as cardboard. He needs to relax, because he is their leader and the room feels everything he feels.

"By the way, Kyle, you didn't even go to college, did you?" you ask, meeting his eyes. There are gasps around the room.

I think it might help if you showed more vulnerability.

"I dropped out," he says, his eyes widening in sudden understanding. "Too many rules to follow."

"So what qualifies you for this job?"

"I wonder that every day," Kyle says, grinning. "I have to learn on the job, fuck up a few times, figure out something new."

The room is calmer now, thrown out of the predictable course of outrage and anger by this back-and-forth. Kyle doesn't know what you're doing, that you're standing by his side but are taller than him in heels, that you're deliberately taking him down a peg to satisfy the mob's desire for a lynching. He just trusts you.

"I think it might be helpful to all of us if you could share why it's important that we diversify the workforce. Not just why it's important for tech—we've all heard the drums—but for *you*."

It's code, and he picks up on it, tells the story that only you know.

"My eldest daughter—I have two—is just starting first grade. She wants to be a scientist. She asked the teacher why people didn't wear their souls on the outside, so that we could end all the problems that came from different colors of skin. The teacher laughed at her and told her that she shouldn't talk about souls if she wanted to be a scientist. Now my daughter thinks she can't be a scientist."

You keep your face neutral so he'll keep talking. It's a start, but he's still checking the boxes of male-ally-bingo, as if it's awkward to put himself into someone else's perspective. It could've been worse, he could've referenced *Lean In*, that magical three-step way for women to achieve everything they want. Be raised by an educated family, go to a pedigree institution to achieve self-confidence, and marry someone who's also educated and rich! So easy.

"I don't think I realized how hard things were for other people because they'd always been so easy for me," Kyle says, admitting his privilege with casual vulnerability. "But now I've got a family and 400 people I feel responsible for, and everything that hurts them hurts me too. I want this to be the kind of place nobody wants to leave. We're not there yet, and I don't have the right answers for how we can get there, but I'm tired of saying goodbye."

This is the moment it all changes.

ONE

THE HEART OF THE MATTER

YOU'RE INVITED (FOR THE first time) to a leadership retreat that includes the brightest technical and organizational minds alive today. A face in the corner looks familiar. With his salt-and-pepper hair Logan is hotter than ever, the rockstar professor from your alma mater who only taught one seminar a year so he could spend a season hacking on various startup ventures. He was awesome. You walk up to tell him how much you've missed him.

Logan remembers you.

He greets you with a warm hug after asking awkwardly if it's okay to hug now that you're no longer his student, reminds you of that one time you said that funny, smart thing and how it totally made his day. Ten years ago. As if that's not enough, he tells you about how he's used you as an example with his peers to bring more women to startups.

You talk for hours. At some point he tells you about a low point in his career, when he'd done everything—*everything*—for his people, but the company was going to go under, and the best he could do was salvage a piece of it so a few people could keep their jobs.

And how one of the people who stayed said, "I can't even pinpoint what it was about how you handled it, but I just don't trust you anymore."

"How did you handle hearing that?"

"Not well," he admits. "It hurts, to know that you can give everything and feel questioned on such a fundamental level about your empathy. To give everything and still feel like you're failing everyone."

"Can we go somewhere?" you ask. "There's something that's been on my mind and I'd like to share but I don't know if I can control my emotions."

He comes with you instantly. Sitting by the fireplace of a seaside resort on a chilly California evening, you tell him about Leo, who told you that you monopolized conversations and couldn't show humility.

"I didn't let him see," you tell Logan, tears filling your eyes. "But it shattered me."

"It's always heartbreaking," Logan says, "to feel questioned about your humanity rather than your competence. To feel that the strengths that are so core to your sense of identity are seen as wicked or weak somehow. I'm analytical, and people think I'm cold. I know someone who's friendly, and people think she's soft."

You wait for the wisdom that will help you bear it all. He's got to know.

"You already do everything that has helped me—yoga, exercise—you can always have a rubber band you snap any time you're on edge..."

Still waiting.

Logan shrugs.

"You'll deal with it," he says casually. "You always do. You were the best student I ever had. Blew my mind actually."

Your heart soars.

TWO

THE DIVINE SECRETS OF THE SISTERHOOD

EVERY MONTH, THE WOMEN leaders of the Valley get together in a SOMA space near Chipotle to talk about whatever is on their minds. Almost all of you are managers, and between the rolling set of ten to fifteen people who show up consistently despite busy schedules, travel, families, and just wanting some time to yourselves, you know there's something significant building here.

Every major company and a few startups are represented. Suman, who bikes forty miles to work every other day, has been working in tech for longer than you've been alive, but she doesn't *ever* pull the "oh, aren't you adorable discovering social justice for the first time" card. Instead she talks about how the lack of women in tech is not a problem but a symptom, and we ought to stop addressing symptoms and troubleshoot the root cause.

"We prize individual endeavors over team contributions in hero culture. It starts in school and continues into the workplace. Maybe the core problem is that we value our individual voices over the voice of the community."

Mohini is subdued these days and says nothing. She spent the last ten years trying to satisfy her parents' desire to see her married to an appropriate Indian boy, but all the appropriate Indian boys have demanded that she stay at home and raise kids or run away screaming when they find out she makes ten times more money than they do.

"I don't think it's that simple," says Alona, who's recently been let go for publishing stories of inaction from her company's HR department and ended up at a higher-paying job running a course in organizational behavior at Stanford. "When I was fired it was not for poking the bear but for talking about poking the bear. Even some of my female former friends think I betrayed the community, that I leaked information by saying anything outside the 'proper' channels even though the proper channels turned out to be shouting into a vacuum."

"They probably think that when you spoke out, it angered the execs who were trying to fix the problem, made them defensive, and set change back a few months," you suggest as an explanation.

"Don't you be 'splainin' for people who aren't even *here*, when they wanted to silence me for saying something uncomfortable. If a company really fucking values *diversity*, they should stop expecting that diverse people will sound like WASP-y white men. Girl, I'm sorry you were colonized at an early age, but if you start blaming the victim instead of accepting her perspective you don't belong here even if it is *your* fucking party!"

You accept the dig and apologize. It's odd, you realize, to be in a room where there are enough women that you might actually have different perspectives on a subject you all care deeply about. So often at your current company you've been called in as The Woman, to proofread job descriptions for inclusive language, to vet a promotion decision, to speak about what needs to be done about diversity. It's refreshing that you and Alona disagree about absolutely everything and love each other anyway.

The conversation moves on to addressing PTSD. So many women end up burned by a bad experience or an inept manager. Even if your heart is in the right place, even if you have the power to change the system, how do you get people to believe you have their best interests at heart?

"One of my reports once asked me why I was asking her to go up for promotion when it wasn't certain," you confess. "She asked me if I was doing it to humiliate her publicly. I don't know how to convince her that I'm not the enemy."

"I have a report who doesn't want to go up for promotion because she's afraid that people will see her as part of the elite and won't like her anymore."

It's Julie, who you've insisted on including in the group, who speaks next. She ditched the philanthropy crap a while ago but got invited to be on the Board of Directors of a growing tech company through a contact she made there. Of course she did, and you're not even jealous because her fiancé cheated on her with some Marina girl that Sam's also been having an affair with, and the Julie who showed up at your door six months ago couldn't even speak. She arrived shivering and barefoot, her feet bleeding from the broken glass of a busted car window, and simply rocked back and forth in your arms for an hour.

"Have you guys told your people that you're in this group, talking through your problems?" Julie asks.

"No, why?"

"Well, when I was at my worst—" she looks at you meaningfully, "—I thought I was completely alone. I'd burned so many bridges, I'd been careless, and... it was the first time I'd ever felt *helpless*. There's something that's odd about being, well, being *us*, being survivors, and feeling helpless. We assume that if we couldn't do it, nobody else could. Because we're strong and we know it. Your reports probably think you're just as powerless as they are. Just as alone. I think it would comfort them to know that you're part of something bigger, something that cuts across companies and cultures. That you're not just one lone woman trying to buck the system."

"True that," Alona says. "We're the divine secrets of the management sisterhood right here."

"Sharing our stories, and our people's stories," says Kiran.

"Because it's not just about bringing up diversity *numbers*," Suman says. "It's about building up enough empathy for others' perspectives so you don't build a product that auto-corrects Suman to Susan every fucking time."

THREE

APHRODITE'S ARROW STRIKES TRUE

You spend two weeks in the hospital. When you wake up you start laughing. Not hysterically, but because it's just so *cliche*. You're an industry leader in your own right now, sponsoring startups that cater to your interests, you get actual *fan-mail*, you're sought after to give commencement speeches in universities, followed on LinkedIn and Twitter by the masses, and your net worth is fast approaching the GDP of Micronesia. You've grown a team out of nothing, with strong women leaders who have revolutionized what it means to work together, to provide high-quality user-facing service, to empathize with customers instead of generating requirements documents.

Getting sick now makes you feel that destiny is a soap opera writer, generating tension where there isn't any so the show can go on for another season.

The doctors have no idea what's wrong. It could be genetic, or chronic stress, or too many long flights, or an autoimmune disorder, or a gluten allergy, or estrogen from years of birth-control, or, or, or…

You do the only thing there is to do: take some time off from work to figure out what you want to do with the rest of your life. You have a month before you need to give the commencement speech at your alma mater, and you spend it at home. Your parents have turned their retirement into a way to help refugees and their children assimilate, and a group of thirty frightened little kids gather in your parents' basement, with their parents admonishing them to behave in front of the VP and not get their clothes dirty.

For the next three weeks you run a workshop on basic computer literacy, ignoring the pinch in your heart when a fourteen-year-old girl asks you if it's illegal to build Android apps because "it's a phone for men" (no woman in their community has the means to purchase a smartphone). They're afraid of breaking the phones you gift them and place them carefully in Ziplock bags instead of using them to hack on stuff.

You do everything in your power to reach them, dressing in torn jeans and lying on your stomach on the floor, propped up on your elbows over a laptop screen as you rediscover the joys of Python with them. They're still scandalized that you're not going to hold an examination.

It's from watching these kids who have been told all their lives that they are wrong, that they are outsiders, that they are different, that they aren't good enough, that they should be afraid, that they should be feared, watching them sparkle with the joy that comes from feeling code come alive beneath their fingers, that you get the words for your commencement speech, the words that send you right back to work.

Today, in Beirut and Cairo, people are using technology to break out of the shackles of war or tradition. Iran's tremendous cultural legacy is being lost to history because we're not able to share information. In rural India and

Pakistan, people are coming online and breaking down borders. This is the most exciting time to be in technology, when we can use it to give voice to previously silent perspectives and connect people who have spent far too long in isolation from each other and new opportunities. The skills you need for tomorrow are not in any particular programming language or field. They are the curiosity to learn and explore, the courage to question your assumptions and pick yourself back up after each failure, the compassion to seek out others' perspectives, and the conviction that we can change the world.

It's not about you. It's never been about you.

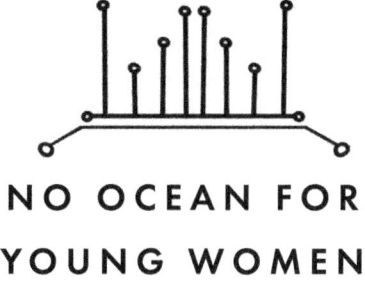

NO OCEAN FOR YOUNG WOMEN

WAVES ARE MERCILESS. THIS is the most important thing you will learn about the ocean, hopefully before you're held down by one for what seems like an eternity, only to have a surfboard crash onto your head when you surface. For years I watched other (taller, stronger) people walk effortlessly through the waves to the lineup where the water was calmer, while I despaired of ever making it there. I stayed in the shallow whitewater, trying to catch the last breath of a broken wave as it fizzled out towards the shore.

Daniella and I are sitting on the beach at Nosara. She is the assistant to surf coach Andrea Diaz, and she's convinced she can help me overcome my fear of the water. It's my first day at the women's surf bootcamp here in Costa Rica.

Daniella asks me, "How do you spot a rip current?"

"The channel where there are no waves," I reply immediately, feeling a creeping dread as I spot the one right in front of us.

She nods and says, "So it's like an elevator, taking you directly to the lineup. If you get caught in one, don't fight it. Wait until it lets you go, then swim away from it and catch a wave back to shore."

It is said so matter-of-factly that it takes me a while to realize the enormity of her words. According to her, a rip current, that confluence of dark swirls that I've seen on warning signs at nearly every beach on the Pacific Coast, is not something to be feared. In fact, fear is the completely natural, yet exactly wrong response, one that can exhaust you or kill you.

But it is one thing to know this, and another entirely to act on it. And I was thirty-five, and at this point had already been struggling with both fear and failure for over a decade.

My first ever surf lesson was at Pacifica. It was a miserable day, cold and with a fog that seeped through to the bone. I was with two of my friends: colleagues who had bought a group lesson. The ocean was choppy, with high winds that created an incessant swirl of seething foam that felt like 1,000 shards of glass. I still remember standing on that stony beach in a rental wetsuit that was damp and reeked of the ocean, being glad of the cowl around my head and my booties. I couldn't feel my toes, and I remember thinking, *My grandmother couldn't even look at Niagara Falls, and I'm trying to surf in the Pacific Ocean.* It felt momentous in that way that things do in your twenties.

"Why are you still standing there?" one of my friends called out to me as he raced into the water. "The longer you wait, the colder you'll get."

He didn't understand that my fear was nearly as deep as the ocean before me. It went back centuries to a peninsula that had remained isolated from the world so that even our gods did not swim, but waited for an army of monkeys to build them a bridge to nearby Lanka. In our mythology, otherwise invincible warriors fell into water and were immediately at risk of drowning unless they were rescued by willing water nymphs who fell in love with them. In the city where I was born, a tsunami had ravaged miles of coastline. My fear wasn't the very rational fear of drowning that can be managed with some good swimming lessons. What struck me when I looked at the Pacific, what still overcomes me when I watch the waves dash themselves against the rocks of Princeton-by-the-Sea, Montara, or Mendocino, is the terror and soul-searing awe that one might feel in the presence of a god.

I didn't stand up on my board that day, although I pushed myself to stand in the water for a little while and tried the pop-up motions I saw others doing. Nobody bothered to put me out of my misery by telling me that you can't really stand on your board unless it's moving, any more than you can balance on your bike when it's stationary. I lied that I'd had fun but was glad to get out of the damp and back to my city life, until the next time another set of guys invited me surfing. It didn't strike me as strange at the time. Gender ratios and working hours in the tech industry meant that most of my colleagues were men, and many of my colleagues were my friends. It also didn't seem strange that I would go back over and over when I wasn't really having fun. I had done a lot of competitive sports growing up, so I knew that you only really started having fun when you had acquired some skill, and I knew the only way to get skilled was to practice.

I took lessons again in Pacifica with my friend Sarah, and in Santa Cruz and in San Diego, wondering why everyone else was able to do things I simply couldn't do. *Lift your chest up*, they said. My chest *was* up, but was also markedly different from that of an average Caucasian woman, and I guess the surf instructors were either unaware or too embarrassed to explain that I would have to work twice as hard to peel myself off the board. *Keep paddling,* I was told. *It gets easier once you're past the waves.* But that point of stillness in the ocean, the lineup where surfers perched to watch the horizon for the next big wave, might as well have been in Hawaii. I paddled until I was panting for breath, until my arms were sore, until a large splash of whitewater knocked me off my board and I thought I was going to drown, only to find that the water was only neck-deep and I had barely made it fifteen feet from shore.

Eventually, I gave up. I became the project manager for surf trips, the person who drove others to beautiful vistas, watched over their valuables and photographed their moments of triumph, ensured that they got out of the water in time for dinner and, on at least one occasion, drove a boyfriend with a broken collarbone to the hospital. But I stayed out of the water. I read and wrote books instead, and the more I saw of the surfing scene the happier I was with my choices. The hard-drinking, pot-smoking beach bum stereotype has given way over the last decade or so to a leaner, more focused kind of masculinity, that of the nerd-jocks who can just as easily explain the physics of wind currents and wave patterns as they can ride them, and who have enough money to invest in new surfboard companies, never mind buy the latest, shortest board on the market and hire a personal coach.

In Southern California, the surfing scene did include several women with perfect arched backs and waist-length hair the color of the sand. If you were dating someone who surfed, you had to surf. One woman told me that her boyfriend simply took her to the lineup and left her there, expecting that she would learn by watching him. It was only when a female friend noticed this that she gave the girl lessons and pulled her away to the beginner beach. But in Northern California, on that colder, more treacherous coastline between San Francisco and Santa Cruz, there were probably more great white sharks than there were women. It wasn't as if there were *no* women in the water. One of them, my former colleague, would charge the waves with her broad forehead completely serene, a tall Nordic goddess who belonged in the ocean and seemed equally at ease running her tech startup as she did when she hit the waves. I watched with awe, but I didn't feel equal to asking her to help me out, any more than I would cold-call my favorite authors and ask them to help me become a better writer.

I told myself I wasn't missing much. In California, there are no seasons, but when the traffic on the 101 slows to a halt with rain, there is a quickening of the heart that can be felt across the valley because it means there is snow in Tahoe. The scent of bay leaf and the promise of fresh berries fills the mountains of Marin, where I would ride horseback, ending my weekends with a canter on the beach. When my commute from San Francisco to the South Bay started to take over three hours a day, I learned to bike the forty miles to work with a small crew of women who didn't mind getting to work a little more slowly and didn't feel the need to take the hillier route via Skyline. It didn't occur to me that I was setting myself up for success on these fronts in very similar ways that I had done in my career. One friend and colleague, a principal engineer, brought discipline and direction to a rag tag crew of cyclists the same way she had when we were on a work project. Another engineer on my team prepared me for a 100-mile bike ride around lake Tahoe by sewing menstrual hygiene pads into my bike shorts, because she knew that tampons just don't make sense to women from certain cultures. The tech sisterhood came together just as much for these things as they did to mentor and advise new graduates joining the workforce.

With the years, I've learned to recognize how much of the course of my career in tech is due to the support, encouragement, advice and sponsorship of others, particularly other women. When I was growing up, in a country where the internet did not (and still doesn't quite) exist, one friend of the family taught me to program in BASIC on an old PC when I was about five or six, while another taught me FoxPro. In high school, my teacher taught me Turing and recommended me for a special summer program when she saw that I loved programming. If it hadn't been for those moments where others believed in me, my own faith would have been sorely shaken in college. There, I was one of perhaps five women in a class of nearly a hundred students, and, during an introductory class, while the professor explained the basics of a while-loop, a guy raised his hand and said dismissively, "This would be much shorter code if you used recursion." I felt the cold, numbing hit of fear (*What's recursion? Was I supposed to know that? Did I miss some pre-work?*) and I told myself to stay calm and focus on the code, not on the other students. At the end of my freshman year, despite a 4.2 GPA, I fell into that vicious whirlpool of having no internship opportunities because I had no prior experience. Then it was a woman professor who gave me a research project, and a woman course administrator who gave me my first job.

To be honest, I've had an easier time than most. For years, that left me blind to what others faced, and quite ignorant of the near-daily scandals now breaking in the news. Having gone to an Ivy League engineering school on a scholarship, I am guilty of carrying at some point or another in my career each of the various misconceptions about why there are so few women in tech. They must not be interested, they don't have the skills, they just prefer to work with people rather than be behind a screen, they leave the industry of their own volition to raise children, always *they*, where other women's experiences were different from mine.

Our minds play tricks on us, and there is no question that mine helped me to distance myself from the struggles of others. I was also the product of an education system that taught me to excel in test-taking environments like interviews, and to see my success as deserved and others' failure as a character flaw. I believed in the existence of a meritocracy because we all shared the same finish line, never questioning whether we had all run the same distance. And the truth is, if my parents hadn't stepped in as educators where teachers had failed, pushing me out of my comfort zone, protecting me from all housework and distractions, and forcing me to take my studies seriously, all while saving up every dime, I probably wouldn't have made it to college at all.

I also have no excuses for being the well-intentioned but terrible manager who tried for years to help people follow *my* career path instead of helping them through theirs. I remember being puzzled at why my hard-won wisdom about having a growth mindset, handling setbacks, leading with optimism, and building the temperament to manage ambiguity, only served to piss off the next generation of women joining tech. Such wisdom is a luxury you can't afford when you feel you're treading water in a rip current and a mile from dry land.

Then, what you may need is a surfboard.

It was my career in tech that taught me to recognize that most men who surf share one story, and most women share another. The stories go like this. A young man walks up to the water for the first time and sees someone his age or younger "totally killing it" and jumps into the water himself. If someone like him can do this, he can too. A young woman walks up to the water and sees a couple of guys doing impossible things in the water, making it look easy. She tries to lift a surfboard and realizes her arms don't really make it all the way around and it's really heavy. She tries to drag it to the water, struggles to paddle, gets flipped over by a wave, and gets out of the water, looking for another woman to teach her.

The stories aren't universal. In Bangladesh, girls as young as ten are learning to surf, crowdfunded thanks to an article in the LA Times. When poverty and the prospect of an early marriage are the future onshore, it is no wonder that these girls fear nothing in the water. And thanks to the pioneering efforts of Marion Poizeau, Shalha Yasini, Easkey Britton and Mona Seraji, men in the Baluchistan province of Iran only ever saw women in the water and believed surfing to be a woman's sport. More and more lines of swimwear now cater to the needs of women who want to fit their "non-standard" bodies into flattering suits, be protected from the sun, be somewhat compliant with religious or cultural norms, *and* perform well on the water.

So what's going wrong on the Pacific coast? First, there are the obvious issues. Even if you got skilled at the actual sport, the water is colder, requiring a wetsuit all the way down the US border. Good surfing conditions are rare, and on the popular waves there is a pecking order in the lineups. Local gangs protect their spots, and outsiders who break the rules can suffer violent consequences. Women who have complained about bullying, sexual and physical harassment, and intimidation on the water have been ignored by local authorities because the laws are of the land, not the water.

Then, there are the more subtle issues, all too familiar to those in tech. Those who have seen the film *Hidden Figures* know that the field of technology was pioneered by and once dominated by women. When it became clear that the field was lucrative, the ways of pushing them out were subtle. Requiring a college degree and recruiting exclusively at certain schools that had historically not welcomed (or accepted) Black students was a good way to change the racial and socio-economic demographics, and citing the thrill of a "startup culture"—long working hours, a college-like scene, hard drinking, and the worship of "brilliant jerks"—was a good way to discourage single parents, older people, and anyone less than confident enough to describe themselves as *brilliant* right at the outset of their career.

Similarly, there is money in surfing, and so the surfing stereotypes are an advertisement geared towards a certain kind of man, one who wishes to break free of the politically correct and neutered workplace and reclaim strength and absolute individual freedom. The language of competence is fundamentally one of violence, and people talk about *charging* the waves, and *ripping, shredding,* or *killing* it out there. To understand how far these images go, you'd have to listen to Laird Hamilton, one of the greatest big-wave surfers in America, tell *TMZ* in an interview, "The biggest, most common reason to be bitten [by a shark] is a woman with her period, which people don't even think about that. Obviously, if a woman has her period, then there's a certain amount of blood in the water." A few media outlets, including *Adventure Sports*, *Popular Science* and *Surfgirl*, took the trouble to debunk this, but when someone spreads a myth (such as, perhaps, a half-baked theory about biological predisposition towards shark attacks or software engineering), it acts like an oil spill. It can only be cleaned up to a certain extent. The damage has already been done.

Yes, you can surf when on your period, which stops when you're in the water. And regardless, you may still want to make sure there are clean shower facilities nearby to prevent urinary tract infections or skin issues. When you surf with other women, you will invariably pick up tips that improve your lifestyle, in or out of the water. The best sunscreens that are non-comedogenic, that will keep your hair color from getting bleached and are also friendly to the disappearing reefs of the world. The value of carrying lubricative tears to wash sand out of the eyes, owning a rash guard, and where to get a two-piece bathing suit that won't fly off when you're hit by a wave. You might learn to use garlic on minor skin irritations or leave a cut onion by your bedside to clear out your sinuses. You may discover the benefits of the post-surf massage and yoga session, and recipes with coconut water, chia seeds and dark chocolate that pull you out of bed for the dawn tide. If the masculine within each of us yearns to surf to face our fears and to battle our insignificance, the feminine yearns to make peace with our vulnerability and achieve harmony with the universe.

The women who come to Nosara to learn from Andrea Diaz are not young. A lawyer, an HR representative, and I are in our thirties and forties. A writer and a photographer for a sports magazine, both in their thirties, are here to capture and describe Andrea's work. A competitive surfer, one of the first women in Costa Rica to take up the sport, Andrea is in her forties and more physically fit than any woman I have met. Still, hers is not the effortless strength of a slender California girl. The years of discipline are visible on her dark skin, and her arms move with the efficient explosiveness of a single mother of three. She teaches us to read the complicated surf charts for wind direction and water currents, and to watch the water to identify the moments when the ocean pauses and allows us to paddle out more efficiently, developing not just our technique but our sense of timing and judgment.

She undoes years of conditioned fear bit by tiny bit, pushing us to do drills in the swimming pool. We get on and off the board a hundred times and retrieve leaves from the bottom of the pool, building lung capacity and the ability to duck underneath the waves. We watch videos to learn technique, seeing that the moves that appeared magical are actually quite doable, and more importantly that the failures, called wipeouts, are usually survivable. I say usually, because I have seen the tombstones at Mavericks and the surreal video of Andrew Cotton being chewed up and spat out of the ocean at Nazaré. As Andrea says, over and over, "Never surf alone," "Never turn your back to the ocean," "Keep the fins away from your face," and "Learn to be comfortable recovering underwater," I realize that what we're actually learning is respect. When we head out to the ocean, I see a four-foot wave about to crash over my head and experience a moment of what feels like wisdom. I remember that the greatest warriors from our mythology have succeeded because in the moments that matter, they display humility. They pray, they kneel before the weapon they would wield, they express gratitude towards their teachers. As the wave comes, I dive deep underwater, kneel into the sand, and bow my head, and experience the remarkable peace and stillness that lies below the barrel of rumbling water overhead.

"If you look down, you'll go down," Andrea says, as we struggle to stay upright on our boards. "Where's your gaze?" That's when I realize that in my career, the hardest transition I have ever had to make is from the frenzy of tactical execution to developing a strategic sense of direction. It is also the transition that so many women struggle to make, crashing hard against that seemingly impossible promotion into the senior ranks. When, as a manager, I have tried to coach people into making it past that point, I've focused on getting *above* the fray, in the same way that I used to try helplessly to get above the waves. *You don't have to do everything yourself*, I'd tell them. *Your job is to figure out where we're going and why.* When they experienced the usual setbacks, had trouble getting heard, or had their ideas dismissed, I told them to focus on the goal and ignore the white noise. But it can still feel impossible, when new tasks and deadlines and expectations arrive without a break. *How am I supposed to plan for next year when I can barely get through this week?* is the common refrain. And just when it seems you've made it, that last, biggest wave—the one that involves personal calamity or deep betrayal—slams into you like an injustice.

So many women in tech burn out treading water in that impact zone, unable to transition to the calmer waters beyond that one, last wave. In the water, I too sense my instincts for tactical response kick into high gear. *The current's pulling me out, the next wave is coming, where's my board, is that other surfer going to hit me, I can't breathe...* and my struggle today is to calm those instincts and remind myself that my only job is to get out of the impact zone. And, importantly, that I'm never alone.

"I'm right here," says Daniella, Andrea's assistant, when I surface after yet again falling into the water gracelessly. "Are you okay? Then get back on your board and paddle. Hold your chest up, head high."

I do as I'm told and make it out to the calm waters where we watch the setting sun and wait for the last green waves of the day. There is nothing in the world quite like the Pacific sunset, and no perspective more perfect than this. This is why I have kept coming back to learn, year after year. Because my love for this world is so much stronger than my fear of it.

A male colleague asked me recently, "Why did you keep going back to surfing? Why didn't you give up? Why did you still go to Costa Rica?" Then he wondered if the problem was with the very question he was asking. He went on, "How many people, when they hear about women in tech's experiences (if they don't first question their validity), just ask why people put up with it? Rather than congratulating them for an achievement. Rather than asking how to remove the BS. They ask them why they've stayed. In tech and in surfing: are women allowed to dream? Or do we just expect them to run away from nightmares?"

He didn't know, and I didn't tell him, that he himself had kept me from leaving tech not once but *twice*. The first time, when I was nearing both boredom and burnout, he offered me a role that spoke to my unique strengths and felt less like trying to paddle against the waves. The second time, when I was struggling to find my feet after a failed project, he helped me find an opportunity where I would be able to grow, and most importantly, built back my flagging self-confidence.

We do ourselves a disservice any time we imagine that there is only one "pipeline" that leads to success, or when we think that success is an objective function of promotions and pay rather than a deep and personal reckoning. I didn't stand up on a surfboard until I was thirty, when I had already worked in tech for nearly a decade. Living in Silicon Valley, I'd done the usual and unusual things—justified a torturous three-hour commute, picked up a taste for ceviche and complex red wines, vacationed horseback across the Wadi Rum in Jordan—and returned to work to confront that question nearly every woman in tech has asked herself: "Is this the life I want to live?"

I went to Nosara for Andrea, for the surf coach who didn't care about my age or my lack of experience, who knew that the most important thing you need when you're scared out there in the water is other people you can trust not to make fun of you when you fail, and to pull you to the surface when you think you're about to drown. This is why, as I now return to the world, to the daily battle against casual and not-at-all-casual sexism, when I must next confront the things and people that mean to unseat and unsettle me, I will know to tell myself and others, *This is just the whitewater. Keep your eyes on the horizon. You're not alone.*

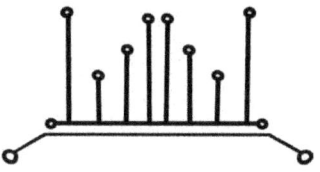

WHY DO YOU BELIEVE YOU'RE A WOMAN?

IF YOU'VE READ MY book, *Driving by Starlight*, you know that I grew up in Saudi Arabia, possibly the most gender-segregated place on the planet. Men wore white thobes, women wore black abayas, and there was nothing but imprisonment or death in-between. You might also have guessed that Leena's experiences of "passing" as a boy are based on my own.

Until I was fourteen (and intermittently after that), I passed as male whenever I felt it would be to my advantage, or would make me feel safer. I continued to do this occasionally as an adult, to the point where a male friend told me, as a compliment, that he saw me as one of the guys, not a girl at all. (Yes, that anecdote in *The Divine Comedy of the Tech Sisterhood*, like so many others, comes from personal experience).

In a country like Saudi Arabia, a child is reminded early and often about their expected gender roles and punished horribly for deviating from them in any way. I was lucky that in this untenable situation, my parents allowed me the fullest freedom I could possibly have. I dressed as a girl for school as was required, and as a boy at all other times. The gym my family went to on evenings and weekends had a gender divider, but it was meant mostly for adults, so my friends and I perched right in the middle, so we could go to either side, whichever had an empty court first.

Although the gender binary permeated every aspect of my existence, I never took it seriously. I played men's parts in school plays because they were more interesting. I grew up with a rich cultural mythology of gender-bending, shapeshifting characters, from a woman who was reborn as a man to exact her vengeance, to a man who became a third gender for a year to hide from his enemies, to a god who took on avatars and switched genders on occasion. And so, even though I was told (usually at parties and weddings when I was forced to dress as one) that I was a girl, that I ought to act less tomboyish and more ladylike, I never quite accepted it.

As I grew older, I started to think about another thing that permeated my life: religion. In Saudi Arabia it was everywhere, and the expectations of religion were often stronger and stranger than the expectations of gender. I was scientific, inquisitive, I wanted proof and got none; instead I got the same message I did about gender: just fake it while people are watching. I asked a Muslim friend at the time whether a truly omniscient God would not know that I was just following the rules but didn't actually believe. She said, and it stuck with me, because a Christian friend in America told me the same thing years later, "If you practice without belief, belief will come to you."

I thought about this a lot, about performing gender and religion, especially as we drove past street signs that said, "Women not allowed" or, "Non-Muslims take detour." In the meantime, I continued to study science, in particular physics and atomic theory. One night, I was staying over at a (male) friend's house and realized the next morning that I had not packed a uniform for school. My friend's well-meaning mother, thinking that any uniform would be better than showing up in yesterday's jeans and T-shirt, packed me off to school in her son's uniform, which fit me perfectly. The science teacher thought to humiliate me in front of the class by asking loudly, "Are you a boy or a girl?" And because I was a smartass, I replied, "I'm an electron. Particle and wave." I was promptly punished (this happened a lot).

It strikes me as strange now that even then I felt the inherent non-dualism at the heart of the universe and the fragility of the colonial dichotomies we impose upon it. Back then I didn't understand why women like my teachers found my masculinity (usually demonstrated through strength and athleticism) so offensive and threatening that they punished it at every turn. But recently I discovered and incredible video of a 1975 interview with Simone de Beauvoir, in which the following lines hit me hard:

Interviewer: "It's women, mothers, who create this discrimination?"

Simone de Beauvoir: "Much of it, yes. Because as the daughters of women, they maintain the tradition… That feminine model is so deeply ingrained in them that they think that a woman who isn't like them is a monster."

The entire interview is fantastic, but the key thing is said at the start without hesitation or caveats:

One is not born a woman; one becomes one. Being a woman is not a natural fact... The more we study the psychology of children, the deeper we delve, the more evident it becomes that baby girls are manufactured to become women.

— SIMONE DE BEAUVOIR.

In other words, practice and belief will come.

Feminist Mona Eltahawy has a project, *Buzzkiller: Memoir in Hair and Revolution*, where she talks about being called "ugly" for having short hair, and being pitied by her own family for being unmarriageable, and others respond with their own stories of bullying and outright harassment over passing as masculine, almost always at the hands of other women. And that's not even getting into the ways in which lesbians' and Black women's femininity is constantly called into question, as if womanhood can only be defined in opposition and in its use to white men.

Why do women, themselves prisoners of the cage, uphold its bars so firmly? Is it jealousy over seeing someone who "passes" having freedoms not accorded to them? I know some of my girlfriends struggled with this, particularly if their own parents didn't allow them to learn how to climb fences or take kickboxing classes. But then surely such jealousy should pass as we become adults, free to do as we please? It doesn't, though. It's almost as if, as adults, the betrayal becomes more evident:

these women spent their entire lives fitting into the mold of womanhood without discomfiting those in power, and are now watching as those hard-won rights are being taken away from all women to punish those who broke the contract of respectability.

And so the questions, "When are you getting settled down? When will you get married? When will you have children?" only get more personal and insistent, and occasionally vicious, as women police each other back into more familiar roles. I have had stressed-out new mothers tell me I must be glad I'm "so free of responsibilities" I'm "basically a man" and women struggling with infertility tell me I "must struggle to find purpose in life."

Age, financial freedom and creative success have allowed me to look past such remarks, but they do make me wonder if they see the stuff of their nightmares in my masculine leanings and child-free existence. Menopause will eventually give us mustaches and take away what's left of the fertility so many women hang onto as the true marker of femininity, the way software engineers hold onto years-stale Computer Science degrees as the marker of their prowess. Perhaps that loss and terror is behind the intensity with which certain older women feel, in the idea of expanding the notion of womanhood to include trans women, the "erasure of the lived reality of women globally." (yes, I can be vicious too).

We are at a moment in time when science has made the barrier between the sexes wholly permeable, so perhaps we can raise the next generation on Ursula Le Guin instead, who told us all the way back in 1969:

> *Consider: There is no division of humanity into strong and weak halves, protective/protected, dominant/submissive, owner/chattel, active/passive. In fact the whole tendency to dualism that pervades human thinking may be found to be lessened, or changed, on Winter... Our entire pattern of socio-sexual interaction is nonexistent here. They cannot play the game. They do not see one another as men or women. This is almost impossible for our imagination to accept. What is the first question we ask about a newborn baby?*
>
> — URSULA K. LE GUIN, THE LEFT HAND OF DARKNESS

As we divide into various forms of us/them, as women are sterilized in camps and the skies in California burn orange because of a fire that started at a gender-reveal party, I wonder what further signs people need to see that binary thinking is killing us.

And yet it is often women who stand most firm against further dissolution of the gender-binary. "But have you experienced a lifetime of abuse and trauma?" they ask. "And if not, how can you suddenly call yourself one of us?" This persists further binary thinking, rooting womanhood in, of all things, trauma. Yes, women are discriminated against, are raped and murdered every day... and thus, the English language permits an equivalent (but incorrect) reading that being a woman is defined, first and foremost, by an assaulter seeing them as such. It's no surprise people make an equivalence between womanhood and trauma. After all, who among us who is seen as a woman has not been attacked for it at one point in time or another? And, uncomfortable as it may be to admit it, was it not such an attack that reminded us or finally made us believe we were women?

"But," you say, "what about periods? Surely that will remain as the Great Differentiator. I can refute all your abstract intellectualism with this bloody tampon."

I'm sure science will find a way to deal with periods (and already has to a great extent). And where science falls short, nature laughs in amusement. Some girls turn into men at puberty instead of getting periods. What is nature if not rapidly adaptable to the needs of the moment, where people wish for nothing more than self-transformation? Transformation is at the heart of so many stories, particularly women's stories:

Women transform because we are hungry. We transform because we're restless, and because we're dangerous. Women transform seeking liberation from domesticity, obscurity, prescribed roles, our own bodies. We transform for fun.

— LARA EHRLICH, *10 SHORT STORIES ABOUT WOMEN'S TRANSFORMATION*

I am reminded of another myth, the one of King Ila, who spent one month as a man and the next as a woman. Ila bore some children as a woman and fathered others as a man, giving rise to the Ailas line: shapeshifters, known as lunar children. Why lunar children? I don't know, but for some reason most shapeshifter lore across cultures has to do with the moon, which changes its appearance over time and reflects the light cast upon it.

The color of the moon, or of any thing, is not an intrinsic property of the object; it is the result of interaction with the light that falls upon it. This smartass learned the difference between phenomena (observed experiences) and noumena (the intrinsic properties of reality) long before *The Matrix* came out, and in response to intrusive questions from family and friends about how much I weighed, I'd respond, "Where? Here or in space?" Observed properties like weight and color have no meaning without context.

I look forward to a day when we no longer have to practice a certain gender to appease family, friends, or doctors who withhold treatment, when women are women because they believe themselves to be and not because of how they are "seen," when children don't come of age in a litany of blood, marginalization, and sexual assault, and when the correct response to "What is your gender?" is not a list of radio buttons but either a blank stare or at least, "When? And in what context?"

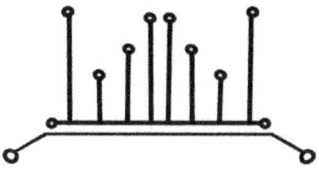

GIVE A WOMAN A MASK, AND SHE'LL TAKE YOU SOMEWHERE NEW

I AM A CHILD of five countries, and can vote in none of them. I speak four languages, and while I write my stories in only one, I can't help but think of them all as mine. So, a few years ago, a certain news story rocked me to my core. A Latina student named Tiffany Martinez submitted a paper and received feedback that the word "hence" was not hers. The story went viral, partially because so many people of color have heard something similar. While every writer yearns to hear feedback, sometimes what you receive can make you angry, or even break your heart. After all, we write to reach out, to be understood, and it hurts to know when it didn't quite work out that way.

Most of the time, editors are well-meaning, and are genuinely trying to help you reach a broader audience. But everyone who gives advice speaks from their own experience, and sometimes what you need is not someone to help you become more like *them*, but to find strength in your own craft. And of course, there will always be those who want to tear you down, who believe they can colonize language itself, exert control over its rules, or exploit it for their own ends. Language is power, as all writers know. Decolonize your writing, and you free your imagination to create worlds better than the one we have.

It isn't easy, though. I have had stories receive entirely different feedback from editors depending on whether I sent them under my real name or my pen-name. It's forced me to come into my own enough to share some of what I've learned so far. Of course, I'm still early in my journey.

I once received feedback on a query that my protagonist must have a last name, not just a first name. But in many countries, particularly in South Asia, surnames only came into use through colonization, and my story was a fantasy world where that colonization hadn't happened. I also received feedback on my use of capital letters, which was no more unusual than other writers, who often use them to talk about some Very Real Thing that isn't necessarily a noun. But of course, it can be hard to tell when feedback is valid, and when you're getting it because someone thinks English might not be your first language.

My breakthrough moment came when I read *The God of Small Things* by Arundhati Roy. Nobody uses capital letters the way she does, and when I saw this sentence I paused, read it twice, spoke it aloud, and then burst into amazed laughter.

"Ammu shook her and told her to Stoppit and she Stoppited."

Could it be? Could a Booker Prize-winning, bestselling author have used the English language to express such a uniquely Indian thought? And if she could do it, why couldn't I? After that, any time I felt annoyed over having my intentions or my fluency questioned, I would imagine an editor's face the first time they saw that sentence, and my mood would lift. Now I can answer, calmly but firmly, *Yes, I meant to do it that way*.

My second lesson would be simplicity. I fell in love with long, periodic sentences, with parataxis and the rhythms of poetry. I was moved to tears by the passage in Martin Luther King's "Letter from Birmingham Jail," that explains why those who have never experienced discrimination find it easy to say, "Wait, change comes slowly." A single sentence, a whole page long, begins with, "But when you have seen vicious mobs lynch your mothers and fathers at will and drown your sisters and brothers at whim…" and goes on, leaving you breathless with a long litany of the indignities of being Black in America, eventually forcing you to despair before ending, "—then you will understand why we find it difficult to wait." That sentence is a perfect union of form and meaning.

But when I used language like that in my fiction, or when I emulated other literary writers' styles, editors would give me feedback like "It's clear you can write, but I just wasn't drawn into the story." I was angry, yes, but eventually realized that my writing wasn't always serving the story I meant to tell. I was writing that way because I felt like an imposter who needed to show these beautiful sentences as credentials to get people to read my story.

Then I read Lee Child. Who, like his character Jack Reacher, has nothing left to prove. And I couldn't stop turning the page. Even when he writes sentences like these that begin with conjunctions. Now, before I write, I say the words, *I don't need to prove myself to anyone.*

Have I conceded too much? Doesn't the language of my colonizer confine me? Of course it does. I once wanted to write a story where the gender of the protagonist remained a mystery till the end, when any assumptions would be shattered. There's no way to do that in third person without making the "reveal" obvious by using "they" from the outset, marking them out as "non-binary" in the Western sense, instead of what I was going for, which was to point out that binary gender was an imperialist construct. Several cultures are more fluid about these things, but the need to classify nature into clean opposites, like good and evil, heroes and villains, black and white, or male and female, has its roots in imperialist attitudes of control and order. The world is wilder than that.

So when I write my characters, I have to be really honest with myself. Am I writing them this way so a Western audience will understand them? Or is this who they really are? That questioning is important, because it is a place of vulnerability, of being comfortable with the ambiguity of the world and human nature, instead of trying to control it. I often ask myself, *Is it still me over here?*

Then I reflect on the strangeness of that question, specifically what it means to "still be me." Authenticity, the way the West understands it today, is tied to individual identity, the borders of our minds and bodies. It started with Descartes' *I think, therefore I am*, which roots authenticity in individual consciousness. That is not how every culture thinks about identity. The

Upanishads speak of a single, indivisible Self manifesting in all beings. In Sanskrit, the phrase *tat tvam asi* (That thou art) expresses that the individual and the Absolute are one, and to claim an individual identity separate from the whole is egoistic delusion.

In these moments, I'm glad of the mask of a penname. I am not bound to the name on my passport, the country of my birth, or the terminology of gender and sexuality that exists today. If Flaubert could write *Madame Bovary*, and if Arthur Golden could write *Memoirs of a Geisha*, why must I be constrained to tell the stories of immigrant assimilation or culture shock? Why would I write purely from and of my own identity? Why would I colonize myself? *I write because I want to be everything, everywhere.*

I can't help this wanderlust. I'm making up for lost time. I grew up in Saudi Arabia before the internet made it possible to explore the world, where women have only recently been permitted to drive and still require the permission of their male guardians to do most things, including work and travel. My novel, *Driving by Starlight*, follows a young girl who dresses as a boy to avoid the religious police from the Committee for the Protection of Virtue and the Prevention of Vice. No, really, that's what it's still called. It was based on my own experiences of being caged in by the world and using fiction to escape.

A strange thing happens when you police so much what people are allowed to do in real life, as every fascist regime eventually discovers: the imagination breaks free. I recently read Anna Burns' *Milkman* and found in it the same irreverence for the rules I have been striving for in my own work. A nameless woman protagonist, with a penchant for reading-while-walking, struggles with a complicated relationship with her maybe-boyfriend and the unwelcome attentions of a stranger. Set during the Troubles in Northern Ireland, where every act is politicized, this story, with its refusal to define and name things, reminded me that as writers, language can and must be our rebellion. The stories we tell, and how we tell them, are how we either perpetuate or break free of the systems of oppression.

When even ordinary things are forbidden, you question everything. When you write, your soul can be genderless and borderless, multilingual and free. Language is not something bound by the systems of the past, but an elemental magic wilder than human ambition and deeper than the sea. Because the true promise of storytelling is adventure, not order. So rather than trying to capture ideas, to colonize them the way we do so much else, including nature and each other, sail among them with curiosity instead. And the next time you hear feedback that tells you the reader is still set in the past, don't let it become a judgment upon your work or your worth. Set yourself free and say to them, *Let me take you somewhere new.*

AFTERWORD

WHY PUBLISH A STORY that's already been published for six years? It started with a friend asking the opposite question after reading *The Divine Comedy of the Tech Sisterhood*: "Why is this not published as a novella instead of being a *Medium* article?"

The answer goes back to January of 2018, when I had not yet published my first novel, *Driving by Starlight*. I spoke to my agent about *The Divine Comedy of the Tech Sisterhood*. Macmillan had optioned my second novel, but only if it was also Young Adult. This story certainly wasn't YA. It was also too short for any publisher to take seriously. Simply put, novellas aren't profitable. She recommended I put the story up on *Medium*, which I did.

I am introverted by nature and didn't grow up with social media. I had no idea how to publicize the story. After *much* encouragement from another friend, I shared it with a former colleague. He loved it and tweeted about it to his thousands of followers. Within days, there were claps, comments, viral threads, and even requests from the *Medium* staff to do an audio version. I was utterly unprepared for this level of public presence and promptly went into my turtle shell. I sometimes remembered to thank people who left comments, but mostly I panicked. Every new comment was a chance for someone to tell me why it was terrible and why, because I wrote it, *I* was terrible.

By contrast, the publication of my first novel, *Driving by Starlight,* was a much quieter affair. I didn't get a book launch—most debut authors don't. I got a few words of critical acclaim, including starred reviews for literary merit from *Kirkus* and *Publishers Weekly*. Yes, there were the occasional one-star reviews that every book gets, but they were off in their corner of Goodreads, not assaulting my inbox.

Now, six years and a second novel later, I realize that "novellas aren't profitable" shouldn't have been an acceptable answer to the writer of this story. Stories aren't about profit; they're about connection. I know now that while the experience of people relating to and loving *The Divine Comedy of the Tech Sisterhood* was overwhelming, it was exactly why I wrote it in the first place—because I was sure there were people out there who needed to hear this story.

I wrote it in a cathartic rage over the course of two weeks in the spring of 2016. Susan Fowler had not yet written her famous blog post about Uber (that would happen only in 2017), and so I felt deeply alone in my experiences as a woman in tech. The only other people who got it were the other women in tech, and together we cry-laughed about our experiences but felt ultimately helpless. When the events described in the chapter "The Woman Whisperer" actually happened to the women on my team, I knew I needed to take a vacation before I said something that would get me in trouble. I took two weeks off, went to a cabin in upstate New York, and wrote the entirety of *The Divine Comedy of the Tech Sisterhood*.

The woman who put the story on Medium in 2018 played within the rules of the game, accepting the world's resistance. She had not yet realized her potential to change things in the workplace, nor did she truly believe her voice had power. I know better now. I accept, with fear and humility, that people look up to me, both at work in tech and as a writer. And the themes in these stories are still relevant, unfortunately. The need to dissociate gender from sex and to dismantle binary thinking has never been more urgent, when abortion rights are being destroyed worldwide and the extremely rich and powerful are dismantling not just DEI initiatives but the basic support structures of society.

I was recently at a conversation with women friends, each of whom worked in a completely different STEM role, and they all had the same stories to share, even a decade later. So I decided to publish this collection of short stories because I wanted to share them with the broader audience my works are getting now, in case there are any people out there who might still need to hear: *You're not alone.* And I wanted there to be a physical collection of my own personal favorite works, the ones I consider my legacy and baton to the next generation, whose feminism is so much braver than my own.

If you enjoyed this book, you might like *Her Golden Coast* by Anat Deracine. When two women with troubled pasts wind up as roommates in early-2000's Silicon Valley, can they heal each other's pain?

ANAT DERACINE is the author of *Driving by Starlight, Her Golden Coast,* and *Algorithms of Betrayal.* She has also written several articles on writing craft, gender, technology, and decolonization, for *Publishers Weekly, Writer's Digest, Mslexia, The Writing Cooperative* and more. She loves engaging with readers and other writers at <u>deracine.substack.com</u>

www.ingramcontent.com/pod-product-compliance
Lightning Source LLC
Chambersburg PA
CBHW032139040426
42449CB00005B/316